Beyond the Evolution vs. Creation Debate

Beyond the Evolution vs. Creation Debate

DENIS O. LAMOUREUX
Professor of Science & Religion

McGahan

Beyond the Evolution vs. Creation Debate

Copyright © 2021 by Denis O. Lamoureux

All rights reserved. No part of this publication may be reproduced, stored in a retrieval system, or transmitted in any form or by any means—electronic, mechanical, photocopy, recording, or any other—except for brief quotations in printed reviews, without the prior permission of the publisher.

All Scripture quotations, unless otherwise indicated, are taken from the Holy Bible, New International Version ® 1973, 1978, 1984, 2011 by Biblica, Inc. ™ Used by permission of Zondervan. All rights reserved worldwide. www.zondervan.com. The "NIV" and "New International Version" are trademarks registered in the United States Patent and Trademark Office by Biblica, Inc. ™

MCGAHAN PUBLISHING HOUSE | LYNCHBURG, TENNESSEE
www.mphbooks.com

Requests for information should be sent to:
info@mphbooks.com

ISBN 978-1-951252-11-3

Contents

CHAPTER 1
Are There Only Two Choices for Origins?   1

CHAPTER 2
Does Nature Point to God?   20

CHAPTER 3
Does the Bible Have Modern Science?   41

CHAPTER 4
What Are the Different Choices for Origins?   70

CHAPTER 5
Are You Ready to Make a Choice?   100

Acknowledgements & Biography   110
Appendix 1: Group Discussion Surveys   111
Appendix 2: Transitional Fossils   112
Appendix Image Credits   118
Notes   121

CHAPTER 1

Are There Only Two Choices for Origins?

Do you believe in evolution or creation? I am often asked this question when people find out that I have written several books on the origins debate. To their surprise, I answer that I believe in *both*. I'm an evolutionist and I'm a creationist.

Most people are quick to tell me that this is impossible. It is not logical to believe in both evolution and creation. I then say to them that I teach at a college and my job title is Professor of Science & Religion. Part of my work is to show students how it is possible to love science and to love God. Another part of my job is to be a scientist who studies fossils and evolution. I often mention that I attend church every Sunday morning and I read my Bible every day in my morning prayers. And my spiritual voyage includes experiencing miracles from God.

At this point in our conversation, some of these people get angry with me. Those who are not religious immediately claim that science proves there is no God. They challenge me to prove that a Creator exists by using scientific evidence. These non-religious people even say to me that *true* scientists are atheists because scientists don't believe in God.

People who are religious challenge me and ask where in the Bible does it say that God created using evolution. They point out that the first chapter of the Bible states the Creator made the entire universe and all living creatures in only six 24-hour days. These religious people then tell me that *true* Christians don't accept evolution.

Now you may think that I get upset when I am told that I can't be both a true scientist and a true Christian for believing in both God and evolution. But I never do. The reason is that there was a time in my life when I said these very same things! I'd often claim that evolution and creation just don't mix, and neither do science and religion. And I would even say this to people in a nasty and aggressive way.

My Story: Trapped in Either/Or Thinking

Let me share a little bit about my personal story. I was raised in a Christian home and I attended a Christian school. I then went on to college to study biology. After taking only one course on evolution during my freshman year, I became an evolutionist. At the same time, I rejected my Christian faith. Professors and older students kept telling me that religion was stupid and intelligent people do not believe in God. The only truth in the world was science. These professors and students had a huge impact on me, and eventually I became an atheist. For me, a world without God meant there was no right or wrong, and I lived my life in a very selfish way.

After college, my faith in God returned. What was the reason? I was so tired of my selfish lifestyle. There was also a small voice in my heart that kept telling me that the way I was treating people was wrong. I started to read the Bible and discovered that being a follower of Jesus was an amazing way to live my life. Returning to God brought a lot of joy and peace into my heart.

I also started to attend church again. It was great to be surrounded by young people my age who cared about doing the right things and living the way God wanted us to live. It was also at church that I met Christians who rejected evolution. They claimed that *true* science proves creation in six 24-hour days is true. For me, this message was important because evolution had destroyed my faith in God while in college. And I quickly became a creationist.

Now I am sure that you have identified my problem. And it's the main problem with the debate over origins today. I assumed that there were only two choices—*either* evolution *or* creation.

Like most Christians and non-religious people, I believed that evolution and science reject God and that this leads to a life with no right or wrong. I also assumed that creation means God made the world in only six 24-hour days. In other words,

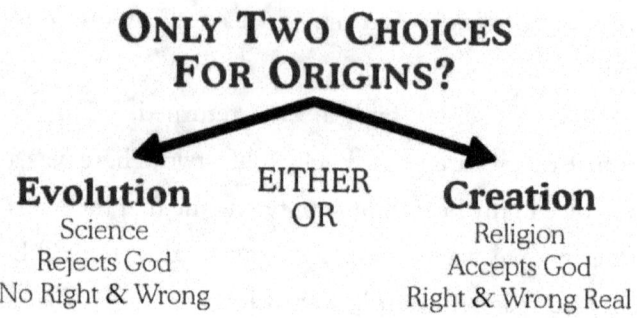

Figure 1-1. The Problem with the Origins Debate: Trapped in Either/Or Thinking

my mind was trapped in either/or thinking. And this forced me into believing there were only two choices for origins. Figure 1-1 presents this problem with the origins debate.

Here is a question for you. Could there be a view of origins in the middle of this diagram between evolution and creation? What about the idea that God created the entire universe and living creatures through evolution? Do you think this is a possibility?

What Does the Word Creation Mean?

Before we can consider the possibility that God created the world using evolution, we need to figure out the meaning of the words evolution and creation. I think the best way to define these words is to look at what they mean to the people who use them in their day-to-day work. Let's begin with the word creation and see how it is used by pastors in churches and professors of religion at colleges.

The word creation is a *religious belief*. It's not a scientific idea or a scientific theory about the origin of the universe and living creatures. Simply defined, creation is the belief that everything that exists in the world was made by the Creator. Therefore, the word creation does not deal with *how* God created stars and planets, plants and animals, and men and women. Instead, someone who is a creationist is a person who simply believes: (1) the Creator exists, and (2) the entire world is his creation.

There are some other religious beliefs that deal with the Creator and the creation.

- Only the Creator existed before anything was created.
- The creation was made out-of-nothing.
- The creation had a beginning and it will have an end.
- The creation is completely dependent on the Creator.
- The Creator can do miracles in his creation whenever he wants to.
- The beauty, complexity, and function of the creation points to the existence of the Creator.
- The creation is very good.
- Men and women are the Creator's greatest creation because we are created in the Image of God.

To conclude, the word creation is a religious belief. For pastors and professors of religion, this word simply refers to everything that the Creator has made, and not to the way he created the universe and living creatures. The belief in creation does not deal with *how* the world was created. Instead, the belief that the world is a creation points us to *who* created it—our Creator.

What Does the Word Evolution Mean?

Let's now look at the word evolution and consider the way scientists define it. Evolution is a *scientific theory*. It simply states that the universe and living creatures were made through the laws of nature and natural processes over billions of years. The theory of evolution does not mention whether these laws and processes in nature were created by God. The reason is because science deals only with the physical world. Science does not deal with God, religion, or the spiritual world.

Now, I need to make a brief comment about the word *theory*. Many Christians assume that a scientific theory is just a hunch or a wild guess. But this is not true. For example, the theory of gravity is accepted by both scientists and non-scientists. I doubt anyone would jump off a building and think it is just a hunch or a wild guess that they would hit the ground! This is also the case with evolution. Today, 98% of scientists accept the theory of evolution and they believe evolution is a scientific fact.[1]

The theory of evolution includes two basic types of evolution: (1) the evolution of the universe beginning from the Big Bang, and (2) the evolution of living creatures including plants, animals, and humans.

The Evolution of the Universe

Scientists have discovered that the universe evolved through the laws of nature and natural processes over billions of years. These laws and processes have made galaxies with stars, planets, and moons. Scientific evidence indicates that the universe began with a MASSIVE explosion about 14 billion years ago. This is called the *Big Bang*.

How did scientists discover this? By looking through telescopes, they noticed that galaxies were moving away from one another. Scientists then reversed the paths of galaxies back in time, and the paths all came together at one single point. This is like taking a video and rewinding it to the beginning. Scientists found that all physical reality was tightly packed into a tiny structure about the size of an atom! It was super hot and super dense. And then it exploded. The universe with its galaxies, stars, planets, and moons slowly emerged through natural laws and processes over billions of years.

Even though science does not deal directly with God and religion, the Big Bang does lead us to ask one important question. Who caused the Big Bang? Did it just happen on its own? I find the idea that the Big Bang occurred without a reason hard to believe. Everyone knows that everything has a cause. So there had to be a cause for the Big Bang. For me, God created the Big Bang.

The Evolution of Living Creatures

The second type of evolution is the evolution of living creatures. This scientific theory explains the origin of plants, animals, and humans through the laws of nature and natural processes. By studying fossils in the rock layers of the earth, scientists have discovered a pattern that shows how living creatures have gradually changed into entirely new creatures over millions of years.

Let me give an example of evolution by using animals with a backbone. These are called *vertebrates*. Fish were the first to appear. Some of them then came on land and evolved into amphibians. Next, some amphibians evolved into reptiles, and then some reptiles evolved into mammals. And at the very end of evolution, some mammals evolved into humans.

Scientists have also discovered that genes slowly change over time. This is the reason living creatures slowly evolve over millions of years into many different types of creatures. These small changes in genes are called *mutations*.

Like the evolution of the universe, the evolution of plants, animals, and humans does not deal directly with God, religion, or the spiritual world. But this second type of evolution makes us ask several important questions.

Why do the genes in living creatures change over time? Is it only luck and blind chance that cause genes to mutate? If this is so, does it mean that the existence of plants and animals and even us is just a fluke of nature? Or have genes been designed by the Creator to change so that living creatures could evolve over time?

I believe God set up the laws of nature so that genes would emerge with the ability to slowly mutate and then create every form of life, including men and women. We are not a fluke or a mistake! We are part of God's plan and purpose. All living creatures evolved exactly the way God wanted them to evolve. In other words, God planned evolution.

Re-Thinking the Relationship between Evolution & Creation

Now that we have introduced the basic meaning for the words evolution and creation, we can explore different relationships between them, as shown in Figure 1-2. Most importantly, this diagram points out a common assumption held by both religious and non-religious people. This is the idea that evolution has no plan and no purpose, and that evolution is caused only by luck and blind chance. Some people believe this. I did too many years ago. But it's not science. Instead, this is a *personal belief* about the scientific theory of evolution.

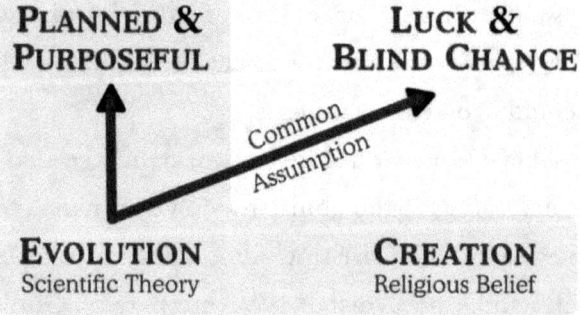

Figure 1-2. Relationships between the Words Evolution & Creation.

The shaded area introduces the idea that evolution has a plan and a purpose. It is possible to accept both the scientific theory of evolution and the religious belief that the world is a creation. This view suggests that the Creator used evolution to create the universe and living creatures.

Let's challenge this common assumption about evolution. And let's also get away from either/or thinking and the idea that there are only two choices for origins. Is it possible that evolution has a plan and a purpose? Maybe God planned evolution so that the universe and living creatures would emerge on their own. And maybe the main purpose of evolution is to create men and women in order that we could have a personal relationship with God. Is this a possibility for you?

This understanding of evolution claims that the entire world *self-assembled* on its own. God did not create using miracles to form each individual star, planet, and moon, or every plant and animal. Instead, God created using the laws of nature and the natural processes of evolution. And all these laws and processes are God's laws and processes. God created them as well.

If you are attracted to this view of origins, then you accept the scientific theory of evolution and the religious belief of creation. This would make you *both* a creationist who believes in a Creator and also an evolutionist who accepts the natural process of evolution. As we will see in Chapter 4, this view of origins is called *evolutionary creation*.

The Similarity between Evolution & Development in the Womb

I am aware that many Christians find the idea of God creating through evolution very troubling. And if this is the case for you, I apologize. But I will ask for your patience. When I first learned about this idea of the Creator using evolution, I didn't like it at all. But I had a friend who helped me by pointing out that there was a similarity between the evolution of living creatures and our own personal development when we were created in our mother's womb, as shown in Figure 1-3.[2] Let me explain.

As a Christian, I believe that God creates each of us by using the laws of nature and natural processes. Scientists call these *developmental* laws and processes. I believe that God made all these laws and processes in order to create everyone. I also love the Bible's use of the poetic language to describe God knitting our bodies together when we were being created. Psalm 139:13-14 states, "For you [God] created my inward parts, you knitted me together in my mother's womb. I praise you because I am amazingly and wonderfully made."[3]

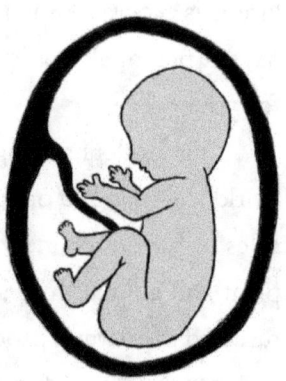

Figure 1-3. Human Development in the Womb

Now, here is my point. I have not met a Christian who thinks that God came out of heaven and used miracles to attach an entire leg or an entire arm to their developing bodies in the womb. And I'm pretty sure you have never met a person who believes that's how we are made. Instead, Christians believe that God creates every person using developmental laws and processes. And God made all these laws and processes. God's creative laws and processes are so incredible that our bodies self-assembled in the womb without any miracles.

So, if we believe that God made each of us by using his developmental laws and processes, then here are some questions to think about.

Is it possible that God created another set of laws and processes in nature that caused all living creatures to self-assemble? Scientists call these *evolutionary* laws and processes. Could it be that instead of coming out of heaven and using miracles to create each creature on earth, God "knitted together" all living creatures by using his natural process of evolution? And if this is true, can we say that God planned and designed evolution, because it has created plants and animals that are "amazingly and wonderfully made?"

For me, our creation in the womb is like the creation of all living creatures on earth. God *ordered* both the developmental and evolutionary laws and processes. God has also *sustained* (maintained) these two natural methods of creation over time. Development in the womb is proof that God uses the laws of nature and natural processes to create. In a similar way, I believe that God used evolutionary laws and processes to create the entire world.

Science & Personal Beliefs in a Complementary Relationship

We now have some new and different ways to think about evolution and creation. This allows us to consider how science and religion can work together in a positive and productive way. Let me propose a relationship between science and our own personal beliefs, as seen in Figure 1-4.

As I said earlier, science deals only with the physical world. It does not deal with God, religion, or the spiritual world. Scientists through their observations and experiments put together theories and laws about the structure,

PERSONAL BELIEFS
Religious or Non-Religious
Ultimate Meaning

Intuition 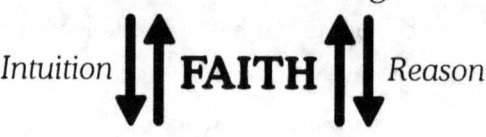 *Reason*

SCIENCE
Laws & Theories
Observations & Experiments

Figure 1-4. Science & Personal Beliefs in a Complementary Relationship

function, and origin of the natural world. And science is incredible! For example, scientists have put rovers on Mars 140 million miles away from earth. And in a very short period of time, they have created vaccines to protect us from the Covid-19 virus.

When we think about all the marvelous discoveries made by scientists today, we cannot help but ask questions about their ultimate meaning. And this is normal. Everyone wonders about the significance of the facts of nature. Therefore, science inspires us to develop our personal beliefs about the physical world. For example, do the universe and living creatures point toward the existence of God and support religious beliefs? Or does nature point away from God and support non-religious beliefs?

Some people come to their personal beliefs quickly through *intuition*. They get a feeling or hunch about the meaning of the world. Other people arrive at their personal beliefs more slowly through *reason*. They use logic and think rationally about the ultimate meaning of the universe and living creatures. Most of us use both intuition and reason to form our personal beliefs as shown in Figure 1-4.

The Upward Step of Faith: From Science to Personal Beliefs

Now here is an idea that needs to be emphasized. After we have used our intuition and reason in thinking about the discoveries of science, we all move from science to our personal beliefs about the natural world. There is no mathematical formula to go from scientific facts to our ultimate beliefs. Instead, everyone takes a *step of faith* as shown in the middle of Figure 1-4. This could also be called an intellectual jump or an intellectual leap. The arrows pointing *upward* in this diagram represent the step of faith that we all make from science to our personal beliefs.

Let me offer an example to explain this upward step of faith. The average cell in our body is about $1/1000^{th}$ of an inch wide. If it is placed on the tip of a pin, we wouldn't be able to see it. In this single cell, there are about two yards of DNA. The amount of information contained in the DNA of this one cell is about the same amount of information found in 6500 books that are each 300 pages long! These are scientific facts about the cell and all scientists accept these facts.

Now here is where it gets interesting. Do these scientific facts point to a God who designed the cell? Or do these same facts about the cell point away from God and the idea that the cell has been designed? Religious people, in taking a step of faith *upward* from science, will come to their personal religious belief that the cell was designed by God.

It is important to point out that non-religious people also take a step of faith *upward* from this same scientific evidence about the cell. However, they come to the personal non-religious belief that the cell is not designed and there is no God.

Of course, you might be asking who is right? Religious people or non-religious people? Ultimate beliefs are very personal, and different people have different beliefs. For me, I have studied the cell scientifically and it blows me away! It clearly looks designed to me and this is one of the reasons why I believe in God. But more importantly, what do you think?

The Downward Step of Faith: From Personal Beliefs to Science
The relationship between science and personal beliefs in Figure 1-4 also has arrows going *downward*. This is something we often miss, and it is sometimes a bit more difficult to understand. So, I'll try to explain this by first saying that we are all raised in a family, we are all educated in different schools, and we are all influenced by our friends and what we see on TV and the internet. Therefore, we are surrounded by all sorts of different beliefs, and these beliefs influence the way we look at the world and think about it.

For example, if you come from a religious family, it is most likely that you have many religious ideas in your mind. These beliefs are like a 'set of glasses' through which you see the world, and they focus your eyes and thinking in a religious direction. When you look at nature, you probably see a lot of design that points to God. And I do too.

On the other hand, if you were raised in a non-religious home and went to non-religious schools and have many non-religious friends, you view the world through a non-religious 'set of glasses.' You might find nature to be amazing, but it does not give you the impression it is designed by God.

So again, you might ask the question, who is right? Religious people or non-religious people? But once more, it comes down to personal beliefs. Different people have different beliefs. Some people believe in God and see the natural world as being designed by him. Other people do not believe in God and do not see design in nature.

But this is the main idea that I want to get across to you. The diagram in Figure 1-4 makes you aware that there is an important relationship between science and personal beliefs. And it is a *two-way relationship*. Your discoveries about nature through science will influence your personal beliefs (upward arrows). And your personal beliefs will influence how you view nature through science (downward arrows).

For me, this two-way relationship between science and my personal religious beliefs is *complementary*. The word complementary means to make something complete, whole, or perfect. For example, in a complementary relationship between two people, each person adds something that the other person does not have. In this way, they strengthen and improve each other.

My science and my religion complete and strengthen each other. The science of evolution shows me *how* God created the world. My personal religious beliefs and the Bible tell me *who* created the world—the God of Christianity.

Therefore, I am no longer trapped in thinking that there are only two choices for origins—*either* evolution *or* creation. I have moved beyond the evolution vs. creation debate and I now see that there is a third choice in the middle. It is very reasonable to believe that God made the universe and living creatures using evolution so that they self-assembled on their own over time. Or to say this another way, I believe *the Creator created an evolving creation.*

CHAPTER 2
Does Nature Point to God?

The universe and living creatures are incredibly beautiful, amazingly complex, and function extremely well. Most men and women throughout time have been deeply impacted by nature. This powerful personal experience has led nearly everyone to believe that the physical world points to a Creator who is like a Magnificent Artist and a Supreme Engineer.

The belief that beauty, complexity, and functionality in nature point to God is known as *intelligent design.* This belief is not restricted to only Christians. Intelligent design is accepted in other religions and by some of the best non-Christian scientists in the world. Anyone looking up at distant galaxies through a telescope, or down into the smallest cells of our body with a microscope, cannot help but wonder if there is an Intelligent Designer who created our spectacular world. The experience of intelligent design in nature is one of the strongest reasons for believing in God.

The term intelligent design is often heard in churches today. Unfortunately, many Christians assume that we must choose between *either* evolution *or* intelligent design. This is like the choice between either evolution or creation, which we examined in the previous chapter. But this approach

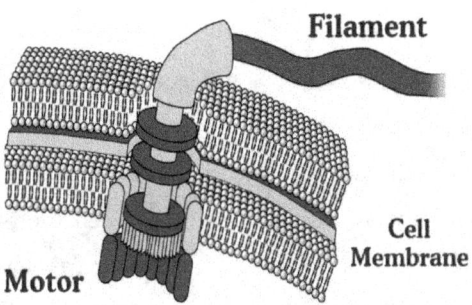

Figure 2-1. The Flagellum of a Bacterial Cell

also traps our mind in either/or thinking. So, let's move away from this common assumption. And let's explore the possibility that evolution might have been intelligently designed by God, and that evolution might even point us toward God.

An example that is often used by Christians to demonstrate intelligent design is the flagellum of a bacteria cell, as shown in Figure 2-1.[4] It has a long thin filament that is attached to a motor in the membrane of the cell. This motor spins the filament at about 1,000 revolutions per minute, and the filament acts like a propeller to move the cell. To be sure, this tiny machine is so spectacular! It certainly looks as if an amazing engineer constructed it.

Here are two questions to think about. Did God create the flagellum in one single moment by using a miracle? Or did God set up the laws of nature and natural processes for the flagellum to evolve and self-assemble on its own?

To answer these questions, we first need to know that scientists have discovered that the different parts that make up the flagellum are also found in the cell membrane of a bacteria cell. And each of these individual parts have different functions. According to these scientists, these different parts self-assembled and created the flagellum. In other words, science claims that the flagellum evolved.

But is it only luck that all these different parts were in the cell membrane having other functions? And is it by blind chance that they came together and formed the flagellum? Or did God set up the laws of nature and natural processes so the flagellum would evolve? For me, the evolution of the flagellum did not happen because of luck and blind chance. The breathtaking complexity and functionality of the flagellum in a bacteria cell *screams out* to me that it has been intelligently designed by God!

We need to ask more questions. By creating the flagellum through evolution, is God trying to get a message across to us? Did he place in the natural world an obvious sign that is pointing to him? Is the Creator speaking to us through the flagellum in the bacteria cell? To answer these questions, we need to explore the various ways that God communicates with us.

How Does God Speak with People?

During my spiritual voyage, I have experienced that God talks to us personally in four different ways: (1) through the voice in our heart, (2) through miracles and answering prayers, (3) through the words of the Bible, and (4) through the natural world and spectacular creations like the flagellum. Let me explain these four ways of how God communicates with us.

First, God has been with me every single day of my life. He speaks to me in my heart all the time. Even as an atheist, God did not give up on me. As I mentioned in the previous chapter, there was a small voice in my heart telling me that my selfish lifestyle was wrong. As I have grown spiritually, I have become better at hearing the voice of God. He has encouraged me, he has challenged me, and most importantly I feel in my heart that God loves me.

Second, there have been times in my life that I have experienced miracles from God. This does not happen all the time. But have you ever had things happen in your life like surprising coincidences that should never have occurred? Most people have had this experience and they believe it is God acting miraculously through these coincidences. God has also answered many of my prayers. You will be encouraged to know that 40% of American scientists believe God answers their prayers in a way that is very noticeable and even tangible.[5] They claim that his answers are much more than just a feeling or hunch.

Third, God speaks to us through the Bible. He inspired the writers of the Bible and he gave them life-changing spiritual truths. These truths are absolutely and totally true. This way of communicating *uses words* and offers specific information about the Creator, the creation, and us. For example, God is love, God is holy, and God is merciful (1 John 4:8; Rev. 4:8; Luke 6:36). The Bible also states that the creation is very good (Gen. 1:31), and that humans are created in the Image of God (Gen. 1:27). I read the Bible every morning and it always encourages me, challenges me, and tells me that God loves me.

Fourth, God calls out to everyone through nature. This way of communicating *does not use words*. It is like a great symphony with only musical instruments. We have all experienced that a symphony stirs in us a wide range of feelings and emotions. Music can create a deep sense of awe, peace, or power. Similarly, the beauty, complexity, and functionality in nature stirs in us the feeling there is something more about the world than just physical things. There is something spiritual. This is the experience of intelligent design. The physical world impacts everyone and loudly declares there is a Creator who is incredibly wonderful and extremely powerful. Intelligent design is a universal music in nature calling us to meet God.

THE BIBLE
Uses Words
Inspired by God
Offers Specific Spiritual Truths
    God is Love, Holy & Merciful
    The Creation is Very Good
    Humans Created in the Image of God

NATURE
Does **NOT** Use Words
Displays Intelligent Design of God
    Beauty, Complexity & Functionality
    Artistic & Engineered
Offers Basic Spiritual Truths
    Nature Points to the Creator
    The Creator is Wonderful & Powerful

Figure 2-2. Two Ways God Speaks to Humans

Figure 2-2 compares how God speaks to us through the Bible and through nature. The Bible uses words and offers us *specific* spiritual truths. The natural world does not use words and offers us *basic* spiritual truths.

For people wanting to meet God, this is what I would suggest. First, begin by experiencing intelligent design in the physical world through its beauty, complexity, and functionality. Nature will speak to you that there is a wonderful and powerful Creator. Second, start reading the Bible to discover who the Creator is. He is a God who loves us more than we can ever imagine. He is also holy and merciful. And God wants to be in a personal relationship with each one of us.

When I explore the natural world through science, the mind-boggling size of the universe and all its marvelous characteristics overwhelm me. To think that the Creator of the world loves me is extremely humbling. Why would he even bother to care about us? But that is who God is. He is the God of Love.

The Bible & Intelligent Design

It is important to mention that the term intelligent design is not found in the Bible. But the idea that nature points to God and offers us some basic spiritual truths about him and the creation appears throughout the Bible. The most well-known biblical passage that deals with intelligent design is Psalm 19:1-4. Let's look at these four verses.

[1] The heavens declare the glory of God;
> the skies proclaim the work of his hands.

[2] Day after day they pour forth speech;
> night after night they reveal knowledge.

[3] They have no speech, they use no words;
> no sound is heard from them.

[4] Yet their voice goes out into all the earth,
> their words to the ends of the world.

Psalm 19:1-4 offers six spiritual truths about the Creator and his creation. These truths align with the belief that the world has been intelligently designed.

First, the creation is active. There are five active verbs in these four short verses. Active verbs describe the actions of someone or something. The heavens "declare," the skies "proclaim," these two heavenly structures "pour forth" and "reveal," and their voice "goes out." Nature powerfully impacts us and shouts out to us that God exists and he is a glorious Creator.

Second, the spiritual truths in nature are understandable. This passage uses four nouns that are associated with intelligent communication: "speech," "knowledge," "voice," and "words." God created the world so that it speaks to us, and God created everyone with an excellent mind to understand the spiritual truths in nature.

Third, the creation's spiritual truths do not use words. Verse 3 says that the heavens "have no speech, *they use no words*; no sound is heard from them." But in the very next verse, it states that there is a "voice" and "words" that go out into the world. This is not a contradiction. It is a way to say that nature certainly talks to us, but actual words are not used. These spiritual truths are like the music of a symphony. By experiencing the beauty, complexity, and functionality of the physical world, it tells us that God is glorious and that he made an intelligently designed world.

Fourth, the spiritual truths calling out from nature never stop. They are heard "day after day" and "night after night" throughout time. This is a way of saying that intelligent design in nature has existed from the very beginning of the world.

Fifth, the spiritual truths in the creation are experienced by everyone and in every place. They travel "into all the earth" and "to the ends of the world." These truths communicated by the creation are not restricted only to Christians and other religious people. These spiritual truths are heard by all men and women everywhere.

Sixth, nature offers spiritual truths about the Creator and the creation. By using wordless beauty, wordless complexity, and wordless functionality, the creation "declares the glory of God" and "proclaims the work of his hands." Nature points to a wonderful and powerful Creator.

I love Psalm 19:1-4. It is one of my favorite passages in the Bible. This psalm explains our experience in nature. It tells us that when we look at the world and get a strong sense or feeling there is someone much greater than us who created it, this is not an illusion, wishful thinking, or just our imagination. It is a real spiritual message calling out to us from nature. This is God speaking to us through his intelligently designed creation. He is telling us that he exists and that he is an incredibly amazing Creator.

Nature & Intelligent Design in a Complementary Relationship

We have now looked at the most important passage in the Bible that deals with intelligent design. Psalm 19:1-4 states that nature offers six spiritual truths. These truths in the physical world are active, understandable, wordless, never stop, impact everyone everywhere, and teach us about the Creator and his creation. Let me suggest a way to understand the relationship between nature and intelligent design, as shown in Figure 2-3. As you can see, this diagram is like Figure 1-4 (page 14) and the complementary relationship we proposed between science and our beliefs.

INTELLIGENT DESIGN
Real or an Illusion
Personal Beliefs

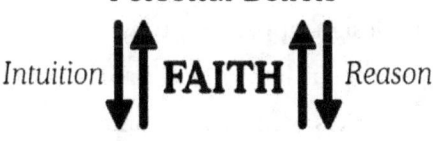

NATURE
Scientific Discoveries
Beauty, Complexity & Functionality
Artistic & Engineered

Figure 2-3. Nature & Intelligent Design in a Complementary Relationship

The relationship between nature and intelligent design is also a *complementary relationship*. That is, it is a two-way relationship. On the one hand, we use our scientific discoveries in the natural world to inform and strengthen our personal beliefs about intelligent design (upward arrows). On the other hand, our personal beliefs about intelligent design are like a set of glasses through which we look at nature and explain the ultimate meaning of our scientific discoveries (downward arrows).

Figure 2-3 also shows that everyone takes a step of faith (or intellectual jump or intellectual leap) in coming to their beliefs about intelligent design. There is a step of faith upward from our experience in nature to our personal beliefs about design. And there is also a step of faith downward from our beliefs about intelligent design that influence how we view and understand nature.

Intelligent design appears throughout the world in a wide variety of artistic and engineered characteristics. Of course, nature strikes different people in different ways. Some are moved more by the breathtaking beauty. Others are hit hard by the complexity and functionality of machine-like structures (e.g., the flagellum in Fig. 2-1 on page 21). Most people experience *both* the artistic and engineered characteristics in the world. In this way, nature features the intelligent designs of an Intelligent Designer who is a Magnificent Artist and a Supreme Engineer. This is my personal experience.

For some people, intelligent design is only an *illusion*. They believe that the experience of design in nature is just our imagination and there is no God. They say that our mind tricks us into believing in design. In other words, the world only gives the *appearance* of being designed. For other people, intelligent design is very *real*, and it points to the existence of God. Most of us today and throughout history have believed that design in nature is a reality. This is my personal belief about intelligent design.

Of course, you might ask the question, who is right? People who believe intelligent design is real. Or people who believe intelligent design is only an illusion. Again, it comes down to personal beliefs. Different people have different beliefs.

In my opinion, I think that our relationship with God influences our belief about intelligent design. Some people have a good relationship with God, others have a bad relationship with God. And still others have no relationship with God. I think everyone will agree that these three different relationships will influence the use of our intuition and reason. In other words, our relationship (or lack of a relationship) with God impacts our thinking.

For example, if someone hates the idea that God exists, has been raised in an atheistic home, and has been educated in non-religious schools, then all the artistic and engineered features in nature will be viewed as only an illusion of intelligent design. Therefore, their personal belief against God will certainly block and blind their view of the clear evidence for intelligent design in nature.

In contrast, someone who loves God and enjoys a good personal relationship with him will surely believe in intelligent design. For this person, the countless artistic and engineered features in nature strengthen their belief in the existence of God. Christians and other religious people undoubtedly see more evidence of design in nature than atheists.

However, I still haven't answered the question you are asking. Who is right? Those who reject intelligent design. Or those who accept intelligent design. For me, intelligent design in nature is very real. As a scientist, I have experienced that the more I study living creatures, greater is the evidence for mind-blowing intelligent design! As well, the great majority of people today and in the past have believed in design. To borrow a term used by lawyers, I believe that the evidence for intelligent design in nature is *beyond a reasonable doubt*.

But more importantly, what do you believe? Does the beauty, complexity, and functionality in the physical world point you to God? Do the artistic and engineered features in nature lead you to believe in intelligent design and an Intelligent Designer?

Intelligent Design in an Evolving & Self-Assembling Creation

Many Christians will find the title of this section surprising (and maybe shocking!). The reason is because we are taught in most churches that we must choose between *either* evolution *or* intelligent design. In other words, we are told that we can't accept both evolution and design. But as we have seen, this either/or thinking traps our mind and we miss the possibility of evolution being intelligently designed by God.

Let me suggest that we move away from this common assumption of choosing between evolution and intelligent design. And let's look at some scientific evidence that suggests evolution points us to God. We will begin by examining the evolution of the universe, and then the evolution of living creatures.

Intelligent Design & the Evolution of the Universe

In the previous chapter, we noted that galaxies are moving away from each other. By rewinding the paths of galaxies back in time, scientists have discovered that all of physical reality was squeezed into a very tiny structure about the size of an atom. Scientists have also found this structure was tremendously hot and tremendously dense. And then the Big Bang happened.

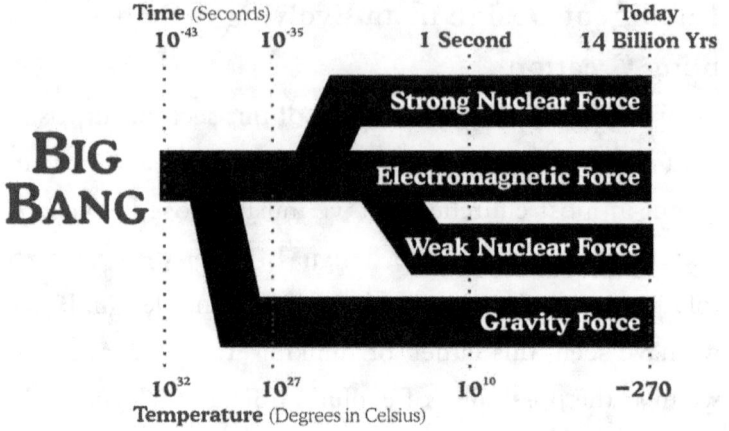

Figure 2-4. The Big Bang & Origin of the Forces of Nature

This massive explosion led to the creation of space and time. As the early universe cooled, physical matter emerged. Then stars, planets, and moons slowly evolved and formed galaxies. I know this might sound weird. But there was no empty space and no time before the Big Bang. Another strange feature is that the forces of nature, like gravity, did not exist either. As Figure 2-4 shows, the four basic forces of the physical world emerged fractions of a second after the Big Bang.[6]

As I mentioned earlier, there must be a cause for the Big Bang. We know that everything in the physical world has a cause. Nothing happens without a cause. For me, God created the Big Bang.

And there is a question we need to ask. Why would the four basic forces of nature emerge just moments after the Big Bang? These four forces are absolutely necessary for our world to function. For example, without gravity everything would be thrown off the earth into space. Is the origin of the four forces of nature only luck and blind chance? I find that hard to believe. It seems more logical to me that God planned the Big Bang and he designed it so the four forces of nature would emerge exactly the way he wanted them to emerge.

There is more scientific evidence about the Big Bang that points to God. Scientists have discovered that the forces of nature are incredibly well-balanced. It is as if someone lined up these forces so the world would emerge. For example, the explosive force of the Big Bang is balanced perfectly against the force of gravity that tries to pull everything back together. If the explosive force had been a bit weaker, the universe would have collapsed back on itself. If it had been slightly stronger, the galaxies would never have formed.

One scientist has calculated the mind-blowing precision between the explosive force and the force of gravity.[7] It is one part in 10^{60}. If we wrote out this number, it is 1 part in 1,000,000,000,000,000,000,000,000,000,000,000,000, 000,000,000,000,000,000. To give you an idea of what this number means, it would be the same accuracy as shooting a bullet and hitting a one-inch target that is

120,000,000,000,000,000,000,000,000 miles away on the other side of the universe!!!

Is this incredible balance between the explosive force of the Big Bang and the re-collapsing force of gravity only luck and blind chance? For me, this scientific evidence points to the intelligent design of God in nature.

The spectacular balancing of these forces is often called *fine-tuning*. It is as if God tuned-up the forces of nature in the way that a car mechanic tunes-up a motor to work at its best. In both cases, an intelligent mind is needed. And I must add that there are many more examples of this spectacular balancing of the forces in the universe. Please see endnote 8 for an excellent book with more stunning examples of the fine-tuning of nature.[8]

Intelligent Design & the Evolution of Living Creatures

Most Christians will be surprised to learn that the evolution of plants and animals also points to an Intelligent Designer. Science has discovered a remarkable pattern with living creatures. As they evolved over time, nearly identical features and identical creatures appeared *independently* from one another. In other words, these features and creatures arose separately on their own. This pattern is known as *convergent evolution*.

To explain convergent evolution, think about evolution to be like a tree, as shown in Figure 2-5. At the bottom, the

Does Nature Point to God? | 37

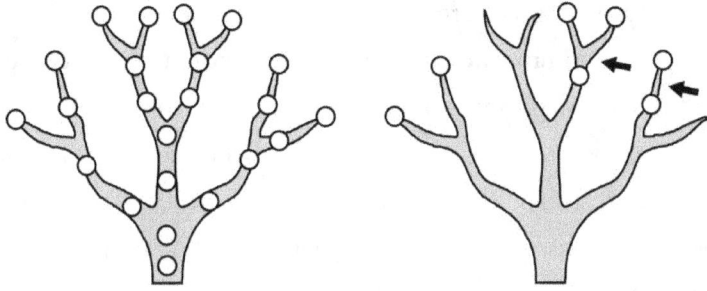

Figure 2-5. Evolutionary Trees
The tree on the left shows a physical feature (represented by a circle) passed on from earlier creatures through evolution. The tree on the right is the pattern of convergent evolution. The same physical feature appears independently and was not passed on from earlier creatures through evolution. But once a feature appears, it can be passed on as shown in two branches on the tree (arrows).

trunk of the tree comes out of the ground and then it spreads out upward into many different branches. Similarly, evolution starts from an ancient creature and then over time evolves and spreads out into many new and different creatures.

If a physical feature was found at the trunk of evolution, then no one would be surprised that this feature would be passed on to all the branches. This would be like passing on a physical characteristic through a family (e.g., eye color). The diagram of the evolutionary tree on the left in Figure 2-5 illustrates this pattern. This evolutionary pattern commonly appears with fossils.

However, convergent evolution occurs on completely different and unrelated branches of the evolutionary tree. As the tree on the right in Figure 2-5 shows, the same physical feature arises on entirely separate branches, and it was not passed on from earlier creatures. In other words, this feature evolved on its own. This convergent evolutionary pattern is also found with fossils.

One of the best examples of convergent evolution is the eye. It has evolved independently over 50 times across a wide range of different animals. These include insects, flat worms, soft bodied marine creatures with shells (molluscs), and animals with backbones (vertebrates). And the camera eye is spectacular. Humans and the octopus have this type of eye. It has an iris, a circular lens, a gel-like fluid inside the eye cavity, and a retina with cells that are sensitive to light. Incredibly, the camera eye evolved independently six different times!

Convergent evolution also occurred with entire creatures. For example, marsupial animals give birth to undeveloped young in pouches, while placental animals are nourished through an umbilical cord in the womb, like us. About 90 million years ago, marsupials and placentals separated from each other. Then they evolved independently on their own. But these two different branches of evolution have strikingly similar creatures. There are marsupial and placental mice, burrowing moles, long-nosed anteaters, saber-toothed cats, and gliding animals like flying squirrels.

The pattern we see with convergent evolution leads us to ask some important questions. Has evolution been planned and programmed so that it would create similar animals and similar characteristics on entirely different branches of the evolutionary tree? Or is the evolution of these same type of creatures and features nothing but a fluke of nature because evolution is driven only by luck and blind chance?

One of the most important scientists in the world who studies convergent evolution claims that if we started evolution again, it would produce basically the same plants and animals that exist today. And he even says that evolution would create human beings just like us. In this way, evolution has been planned and programmed to create the living creatures that now exist. Please see endnote 9 for this scientist's brilliant book on convergent evolution with over 400 examples of similar features and creatures that evolved on entirely different branches of the evolutionary tree.[9]

For me, convergent evolution is scientific evidence that indicates evolution is not the result of only luck and blind chance. The appearance of similar features and creatures on separate and independent branches of the evolutionary tree (Fig. 2-5 the tree on right) is evidence that the Creator planned and programmed the natural process of evolution.

Let's bring this chapter to a conclusion. When I look at nature, it shouts out to me that there is a God. The beauty of oceans and mountains strike me as the artwork of a Magnificent Artist. The machine-like complexity and functionality of plants and animals make me think that these were constructed by a Supreme Engineer. And when I think about the Big Bang and the evolution of living creatures, they too "declare the glory of God" and "proclaim the work of his hands" (Ps. 19:1). It is obvious to me that the Creator speaks to us through his marvelous self-assembling creation. Indeed, nature points us to God.

CHAPTER 3
Does the Bible Have Modern Science?

When I mention to Christians that I believe in the Bible and that I also accept evolution, some of them get aggressive with me. They are quick to tell me that *real* Christians cannot accept evolution. Then they challenge me, "Where in the Bible does it say that evolution is true?" And I am warned, "The Bible states that God created the universe and living creatures in only six 24-hour days. If you are going to be a *real* Christian, then you need to believe this. *Your faith depends on it!*"

You may think that I get upset with these Christians. But as mentioned in Chapter 1, I never do. The reason is that there was a time when I said all these same things to Christians who accepted evolution. In fact, I admire Christians who criticize me, because I know they love God and they love the Bible. They see evolution as a threat to Christianity. These Christians believe they are defending and protecting what God has said in the Bible. And I respect them for that.

Sometimes I sense that I might be able to have a productive conversation with Christians who disagree with me about evolution and the Bible. I always begin by talking about spiritual experiences that we share. For example, most Christians will say they have experienced that God meets us exactly where we happen to be in our life. In other words, God comes down to our level when he speaks to us.

Then I ask, "Is it possible that God came down to the level of the ancient writers of the Bible when he spoke to them, in the same way he speaks to us today?" I have discovered that most Christians have never thought about this idea. But they realize this is reasonable. After all, when we pray to God, he does descend to speak with us so we can understand him. Think about it. He is the Creator and we are only human creatures! God needs to come down to our level for him to talk with us.

Since nearly all Christians agree that God meets us wherever we are in our life, I then suggest a new and challenging idea. And I do this by carefully asking three questions. Is it possible that when God inspired the ancient writers of the Bible, he descended to their level and spoke to them by using their ancient understanding of nature? If this is true, did God allow the ancient writers to use the science-of-the-day in the Bible? To be more precise, is it possible that God talked to them using *ancient science*?

However, most Christians are not at all comfortable with the idea that the Bible has ancient science. They quickly ask, "Are you saying that God LIED in the Bible?" I admit to them that this is a fair question. But I make it very clear that God does *not* lie in the Bible. I then show them two Bible verses. Titus 1:2 states that God "does not lie" and Hebrews 6:18 says "it is impossible for God to lie." Lying requires someone to be dishonest, and the God of the Bible is not a God of dishonesty.

Therefore, God did not lie in the Bible if he allowed the ancient biblical writers to use the only science that they understood. Instead, he came down to their level so they could understand what he is saying to them about the natural world. Of course, God could have inspired the author of the first verse of the Bible to write: "In the beginning God created the universe through the Big Bang and living creatures through evolution." But would any ancient person have understood this verse? I don't think so.

Let me give an example to further explain this idea of God coming down to the level of the ancient writers of the Bible. When a 4-year-old child asks about where babies come from, do parents talk about the modern scientific details of sexual reproduction? Of course not. Instead, they get down on their knees and use simple words and simple ideas that a 4-year-old can understand. And most importantly, they offer the spiritual truth that God creates babies and that every baby is a gift from God to their parents.

Most Christians will agree that the main purpose of the Bible is to give us spiritual truths about who God is and who we are. Take the first chapter of the Bible—the Genesis 1 account of creation. What are the two main spiritual messages? Christians would say: (1) God is the Creator of the world, and (2) humans have been created in the Image of God. Genesis 1 also mentions that birds were created before land animals. But is that a spiritual truth? Will that change your life and bring you closer to God? I don't think so.

It must be pointed out that there is a good reason why so many Christians find it difficult to accept the idea that the Bible has an ancient science. Our churches, Sunday schools, and Christian schools teach us that the Bible has some true scientific facts about the natural world. In other words, most Christians believe the Bible has modern science. Many years ago, I believed that. But is this true? Does the Bible have modern science?

In this chapter, we will attempt to answer these questions. I'll begin by looking at passages in the Bible that deal with the heavens and astronomy. Next, biblical passages about the earth and geography are examined. By the end of this chapter, you will be able to determine if the Bible has modern science or ancient science.

Looking at Nature through Ancient Eyes

When I was in college, I studied the first chapters of the Bible. One of the best lessons that my professors taught me was that we need to read the Bible the way a person in ancient times read it. This makes sense because the Bible was written a long time ago. So, when the biblical writers talk about the natural world, we need to look at nature through their ancient eyes.

To explore how the writers of the Bible understood the natural world, let's think about the sun. I am sure most of you know that ancient people believed that the sun *literally* and *actually* moved across the sky every day. They did not know that the earth rotates on its axis and that this causes day and night. Instead, ancient people thought that the sun literally rises in the east and then literally sets in the west.

From the point of view of an ancient person, the idea that the sun actually moves across the sky daily is very reasonable. They did not have modern scientific instruments like telescopes to show them that the earth rotates. They only had their naked eyes. And they saw the sun moving each day. As well, they didn't feel the earth moving. Therefore, when ancient people looked at the natural world, they had an ancient point of view. Or stated in another way, they had an *ancient perspective of nature*.

This ancient understanding of the literal and actual movement of the sun appears in the Bible. Ecclesiastes 1:5 states, "The *sun rises* and the *sun sets*, and hurries back to where it *rises*." Similarly, Psalm 19:4-6 says, "In the heavens God has pitched a tent for the sun. It is like a bridegroom coming out of his chamber, like a champion rejoicing to run his course. It *rises* at one end of the heavens and makes its circuit to the other." And Psalm 113:3 claims, "From the *rising* of the sun to the place where it *sets*, the name of the Lord is to be praised."

Now, there are some Christians who will say that I am misunderstanding these biblical passages. They argue that the terms "sunrise" and "sunset" are figurative language and are not to be taken literally. These Christians often add, "Check out any weather app today and you will find the times for sunrise and sunset. But everyone knows that the sun does not actually rise or set. Therefore, verses in the Bible referring to the sun moving across the sky should be read figuratively and not literally."

This seems to be a reasonable argument. But there is a question we need to ask. Everyone knows that in ancient times people believed the sun literally moved across the sky every day. But when did scientists discover this was not true?

The answer is around the 1600s, when the first telescopes were made. The famous astronomer Galileo showed that the earth rotates on its axis daily. He also made us realize that movement of the sun across the sky is only an *appearance*. It's just a visual effect. Therefore, it was only after the 1600s that people began to use the terms sunrise and sunset figuratively and not literally.

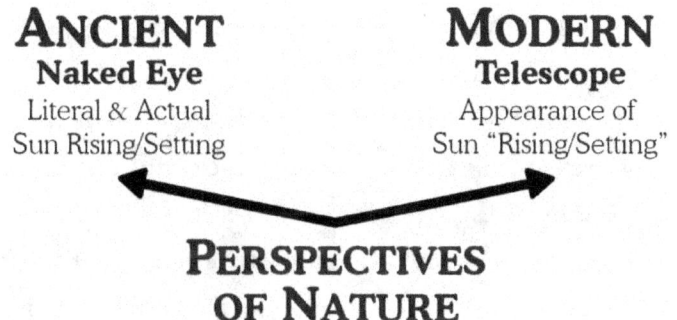

Figure 3-1. The Ancient & Modern Perspectives of Nature

This fact of history is important for reading the Bible. We know that the Bible was written from about 1400 BC to AD 100. This period is hundreds of years before the scientific discovery in the 1600s that the sun does not literally move across the sky every day. So, when the biblical writers said that the sun moves, they meant that the sun literally and actually moves. And this ancient science was the science-of-the-day in ancient times.

Figure 3-1 presents the two basic ways of looking at the natural world. Ancient people had an *ancient perspective of nature*. Their point of view was based on what they saw through their naked eyes. They did not have our modern scientific instruments. Ancient men and women believed that the sun literally and actually moves across the sky each day. The writers of the Bible had this ancient perspective of the sun. When they said that the sun rises and the sun sets, they meant it literally.

In contrast, we have a *modern perspective of nature*. We live during a remarkable time with fantastic scientific instruments like powerful telescopes. When we see the sun "moving" across the sky, we know it is only an appearance. It's just a visual effect caused by the rotation of the earth. Therefore, we use the terms sunrise and sunset figuratively.

It's important that we do not confuse the ancient perspective of nature with our modern perspective of nature. When we read passages about the natural world in the Bible, we need to read them through the ancient point of view of the biblical writers. In other words, we must look at nature through their ancient eyes so that we can better understand what they wrote in the Bible.

The Relationship between Ancient Science & Spiritual Truths

I suspect that many of you are probably asking the question, what are we to do with the idea that the Bible has an ancient science? This is a great question. Figure 3-2 introduces a way to relate the ancient science in the Bible and the spiritual truths that God has given us in the Bible. Let me explain this diagram.

First, I think every Christian will agree that the main purpose of the Bible is to teach us spiritual truths. God inspired

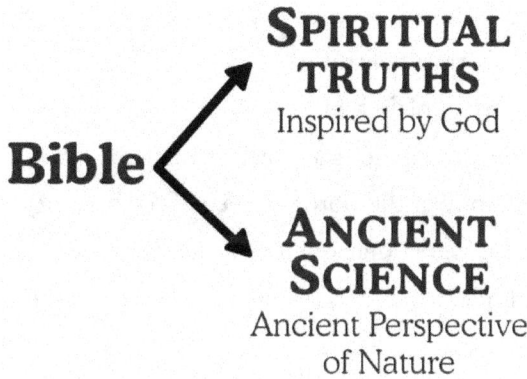

Figure 3-2. The Relationship between Ancient Science & Spiritual Truths

the writers of the Bible to write down these truths. And these spiritual truths are *absolutely true*. They have powerfully impacted the lives of men and women throughout time. They tell us who God is and who we are. They also bring us joy, comfort, and meaning. These spiritual truths help us develop our personal relationship with God.

Second, when the Bible mentions the natural world, it uses ancient science. This was the best science-of-the-day in ancient times. As we saw in Figure 3-1, this ancient science is based on an *ancient perspective of nature*. Ancient people did not have modern scientific instruments like telescopes. They only had their natural senses, like the naked eye. Therefore, God came down to the level of the ancient writers of the Bible and used their ancient science like a vessel or container to transport his spiritual truths.

I need to explain this a bit more. Think of a water bottle. Does it matter whether the bottle is made of plastic, or glass, or metal? No. But the bottle is needed for holding the water. And it's the water that quenches our thirst, not the bottle. This is similar to the Bible when it mentions the natural world. The ancient science is like a bottle. It holds the spiritual truths. And it is these truths from God that quench our spiritual thirst, not the ancient science.

Let's look at a passage in the Bible and apply this relationship between ancient science and spiritual truths seen in Figure 3-2. Earlier we saw that Psalm 19:4-6 states, "In the heavens God has pitched a tent for the sun. It is like a bridegroom *coming out* of his chamber, like a champion rejoicing to *run* his course. It *rises* at one end of the heavens and *makes its circuit* to the other."

It is obvious that God allowed the biblical writer of this psalm to use the ancient scientific idea that the sun moves across the sky every day. In comparing the sun to a bridegroom and a champion, this writer uses four verbs to describe the sun moving—the sun comes out, runs its course, it rises, and makes its circuit. Is the purpose of Psalm 19:4-6 to give us scientific facts about how the sun moves? I don't think so.

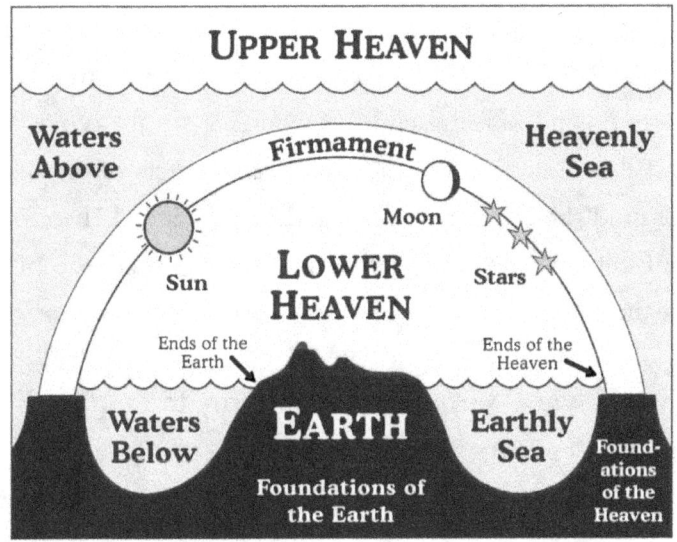

Figure 3-3. The Ancient Understanding of the Structure of the Heaven & the Earth

The purpose of Psalm 19:1-4 is to give us the spiritual truth that God is the Creator of the sun and the heavens.

Now, did you notice another ancient scientific idea in Psalm 19:4-6? It's easy to miss. The biblical writer compares the heavens to a tent. And what is the shape of a tent? It has a canopy overhead that is like a dome. From the perspective of the naked eye, isn't this what the sky looks like? As Figure 3-3 shows, this was the science-of-the-day in the ancient world.

To conclude, the purpose of the Bible is to give us spiritual truths, not scientific facts. In the same way that God comes down to meet each of us wherever we are in our life, God also came down to the level of the writers of the Bible and used their ancient understanding of nature. Therefore, the Bible does not have modern science. Instead, the Bible has ancient science.

The Heavens & Ancient Astronomy in the Bible

Let's now look at some more passages in the Bible that deal with the heavens. I'll try to show you that there is an ancient astronomy, like the ancient scientific idea that the sun moves across the sky each day.

It is important to remember that when we read passages about nature in the Bible, we need to look at nature through the eyes of ancient men and women. In other words, we need to use their ancient perspective of nature (Fig. 3-1). If we do this, we will discover that the Bible has an ancient understanding of the structure of the heavens and the earth, as shown in Figure 3-3.

The Firmament & the Heavenly Sea

One of the most surprising ancient scientific ideas that I discovered in the Bible is that the dome of the sky was believed to be solid and hard. As well, the solid dome held up a heavenly sea of water. This ancient astronomy appears in the first chapter of the Bible. The hard dome was called the *firmament*. On the second day of creation, Genesis 1:6-8 states:

> 6 God said, "Let there be a firmament between the waters to separate waters from waters." 7 So God made the firmament and separated the waters below the firmament from the waters above the firmament. And it was so. 8 God called the firmament 'heaven.'

The Bible was originally written in the Hebrew language. In this passage, the original Hebrew word *raqia* is translated into the English word firmament, and it means the hard dome of heaven. Ancient people thought it was like an inverted bowl. Why did they believe this? Well, think about it. When they looked up at the sky, they saw a massive blue dome. For them to believe there was a solid and clear structure that held up a heavenly sea of water made perfect sense. It's what they saw with their naked eyes.

Figure 3-3 labels the domed firmament in heaven. This diagram also shows the heavenly sea. It is called the *waters above the firmament* in Genesis 1:7. And in this same verse, the *waters below the firmament* refer to the earthly sea. I am sure you will agree that this ancient understanding of the structure of the heavens is not modern science. This is ancient astronomy. But the spiritual truth in Genesis 1:6-8 is obvious—God created the sky.

The Sun, Moon & Stars in the Firmament

The fourth day of creation in Genesis 1 also presents an ancient understanding of astronomy. The sun, moon, and stars are placed in the dome of the heavenly firmament. Genesis 1:14-17 states:

> [14] God said, "Let there be lights in the firmament of the heaven to separate the day from the night, and let them serve as signs to mark sacred times, and days and years, [15] and let them be lights in the firmament of the heaven to give light on the earth." And it was so. [16] God made two great lights—the greater light to govern the day [the sun] and the lesser light to govern the night [the moon]. He also made the stars. [17] God set them in the firmament of the heaven to give light on earth.

So, why did the ancient writers of the Bible believe in this ancient astronomy? From the point of view of the naked eye, this is what it looks like. The sun, moon, and stars are right up against the dome of the sky. This gives the appearance that these heavenly lights are set in the surface of the firmament, in front of the heavenly sea. The ancient perspective of the heavens is quite logical. If we had lived at that time, we would have believed this.

Figure 3-3 shows the positions of the sun, moon, and stars in the heavenly dome of the firmament. Of course, this is not the actual structure of the heavens or the position of these heavenly bodies in outer space. This is ancient astronomy, not modern astronomy. But Genesis 1:14-17 offers a clear spiritual truth—God is the Creator of the sun, moon, and stars.

The Ends & Foundations of the Heaven

Now that you have a basic knowledge of the ancient astronomy in the Bible, we can return to Psalm 19:4-6 and have a better understanding of it. "In the heavens God has pitched a tent for the sun. It is like a bridegroom coming out of his chamber, like a champion rejoicing to run his course. It rises [1] at one end of the heavens [2] and makes its circuit [3] to the other."

As Figure 3-3 shows, the ends of the heavens are at the horizon, where the firmament comes to an end. In this passage the sun rises at one end of the firmament, moves across the circular dome of the firmament overhead, and then sets at the other end of the firmament. The ancient astronomy in Psalm 19:4-6 is reasonable if the heavens are viewed from the perspective of the naked eye.

Ancient people also saw that the firmament did not move. They were logical in assuming that the firmament was set on a solid base, like the foundations of a building. For example, 2 Samuel 22:8 states, "The earth trembled and quaked, the *foundations of the heavens* shook; they trembled because God was angry." The writers of the Bible experienced earthquakes. The idea that God could shake the foundations of the earth and the foundations of the heaven made perfect sense to them. Figure 3-3 identifies these two types of foundations. Obviously, this is not modern science.

The passages above that refer to the ends and the foundations of the heavens offer important spiritual truths. Psalm 19 says that God created the heavens and the sun. In 2 Samuel 22, the spiritual message is that God is in total control of the world and he can shake it whenever he wants. The ancient astronomy in these biblical passages is a vessel that transports these spiritual truths about God.

The Upper & the Lower Heavens

Ancient people believed that the heavens had two main regions. The upper heaven was the dwelling place of God and the angels. The lower heaven included the air, the heavenly sea, and the firmament with the sun, moon, and stars. The Bible presents this ancient two-part structure of the heavens, as shown in Figure 3-3.

The upper heaven appears in Deuteronomy 26:15. The nation of Israel prayed to God, "Look down from heaven, your holy dwelling place, and bless your people Israel." Nehemiah 9:6 refers to both the upper and lower heavens. The Israelites praise God, "You made [1] the heaven, even [2] the highest heaven, and all their starry host." In this verse, the word heaven refers to the lower heaven, and the term highest heaven means the upper heaven.[10]

The lower heaven includes the firmament. For example, on the second day of creation in Genesis 1:7, God "called the firmament 'heaven.'" The air is also part of the lower heaven. This is where birds fly and clouds appear. Genesis 2:19-20 and Matthew 6:26 both use the term the "birds of the heaven." And Daniel 7:14 and Mark 14:62 mention the "clouds of heaven." Finally, the lower heaven includes the heavenly sea. Psalm 148:4 and Jeremiah 10:13 both refer to the "waters in the heaven." Together, these biblical verses present an ancient science, not modern science.

The use of ancient astronomy in the Bible allowed God to speak to ancient people about the natural world in a way they could understand. God came down to their level to say that he was the Creator of the heavens. And God even answers our prayers and blesses us from heaven. These are spiritual truths that are transported by the ancient understanding of astronomy in the Bible.

The Earth & Ancient Geography in the Bible

In the previous section, we discovered that the Bible has an ancient astronomy. It is only reasonable and consistent to suggest that the Bible also has an ancient understanding of the structure of the earth. In other words, there is an ancient geography in the Bible, as shown in Figure 3-3.

The Immovable Earth & the Foundations of the Earth

People in ancient times believed that the earth did not move. This was a reasonable idea from their ancient perspective of nature. Even today, we do not feel the earth spinning at 1,000 miles per hour and moving around the sun at 70,000 mph. The experience that the earth does not move was so powerful that people believed the earth was immovable until the 1600s. It was during that time Galileo offered scientific evidence to show that the earth moves.

The Bible clearly presents the earth as immovable. For example, Psalm 93:1 says, "The world is firmly established; it cannot be moved." The writers of the Bible often used the engineering term foundations to explain the stability of the earth. As Psalm 104:5 states, "God set the earth on its foundations; it can never be moved." Figure 3-3 identifies the ancient scientific idea of the foundations of the earth.

Like the ancient astronomy in the Bible, the geography in the Bible reflects an ancient understanding of the structure of the earth. The ideas that the earth did not move and that it was set on foundations are ancient ideas. This is ancient science and not modern science. But more importantly, the ancient geography is like a container that transports the spiritual truth that God created our wonderful home, the earth.

The Circular Sea & the Circle of the Earth

These next two ancient geographical features will strike you as very odd. Ancient people in the Middle East believed the earth was surrounded by a circular sea. They also thought that the earth was an island and that its shoreline was the shape of a circle. To understand these ancient ideas, we need to look at the world through ancient eyes and think like an ancient person.

Two experiences in nature led ancient people to believe that a circular sea surrounded a circle-shaped earth: (1) When they looked at the horizon, it gave the impression that the entire world was enclosed inside a circular boundary. As we have seen, ancient people believed the firmament was like an inverted bowl overhead. (2) When ancient people traveled in almost any direction in the Middle East, they came

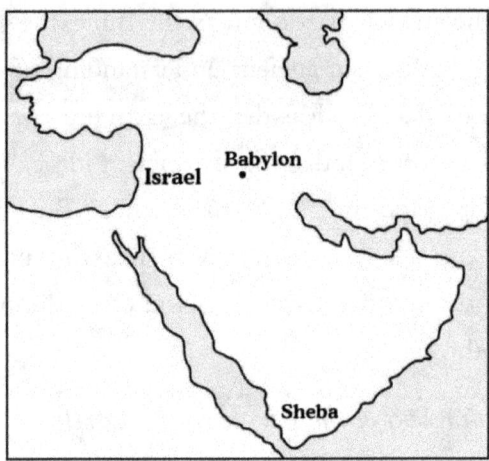

Figure 3-4. Geography of the Middle East

to the shoreline of a large body of water. Figure 3-4 is a map of this region and shows the land is surrounded mostly by seas.

This ancient geography appears in a map of the world that was made around 600 BC. It is called *The Babylonian World Map*, as shown in Figure 3-5.[11] The city of Babylon was thought to be at the center of the earth. A circular sea surrounded the earth. The outer boundary of the sea was at the horizon where the firmament ends. The earth was believed to be in the middle of the world. And the shoreline of the earth was thought to be a circle. This was the best geography-of-the-day.

Figure 3-5. The Babylonian World Map

The Bible includes this ancient geography. Isaiah 40:22 says, "God sits enthroned above the *circle of the earth*, and its people are like grasshoppers. God stretches out the heavens like a canopy, and spreads them out like a tent to live in." If we consider the ancient understanding of the universe in Figure 3-3, then this verse makes perfect sense. Ancient people believed heaven was like the domed canopy of a tent, and the earth was like the flat floor of a tent. Therefore, the circle of the earth in Isaiah 40:22 refers to the circular shoreline of the earth where it meets the circular sea.

The ideas of the circular sea and the circle of the earth are ancient science, not modern science. This is ancient geography. But more importantly, Isaiah 40:22 offers spiritual truths. God is the Creator of the world and he made it specifically for us to live in. God is also the King of the universe. He sits on his throne in heaven and rules over the earth.

The Ends of the Earth

In the previous section, we saw that ancient people traveling in most directions in the Middle East eventually came to the end of dry land. They believed this was the shore of a circular sea that surrounded a circular earth. Therefore, it was quite logical for them to call this shoreline the *ends of the earth*. This term appears nearly fifty times in the Bible. Let's look at a well-known example.

In Matthew 12:42, Jesus states, "The Queen of the South will rise at the judgment with this generation and condemn it; for she came from the *ends of the earth* to listen to Solomon's wisdom, and now something greater than Solomon is here." Jesus was referring to the Queen of Sheba. Her country was in the southwest corner of Arabia, and it was bordered to the south and west by seas, as shown in Figure 3-4. From an ancient perspective of nature, Sheba was at the ends of the earth.

I think most of you will agree that Jesus was not trying to teach geography in Matthew 12:42! Jesus was talking about himself. He was saying that his teaching is greater than the wisdom of Solomon. Jesus was also telling us that there will be a Final Judgment at the end time. But the Good News is Jesus shows the way to be with God forever.

The Flat Earth

Most of us are aware that ancient people believed the earth was flat. This ancient idea is reasonable. Any person looking out from an elevated position like the top of a mountain will see that the earth appears to be a level plain bordered by a flat circular horizon. In other words, the belief in a flat earth was based on an ancient perspective of nature.

The flat earth is implied in Matthew 4:1-11, when Jesus is tempted by the devil. "Again, the devil took him [Jesus] to a very high mountain and showed him *all* the kingdoms of the world and their splendor. 'All this I will give you,' he [the devil] said, 'if you will bow down and worship me.'" However, we know that there were great kingdoms in China and Central America during the days of Jesus. It was not possible for Jesus to see "all the kingdoms of the world" from any mountain on our spherical earth. Therefore, Matthew 4:1-11 implies the earth is flat.

I must also point out that the Bible mentions the word earth nearly 2800 times. But not once does the Bible say that the earth is a sphere. Think about this. If God wanted to teach us modern scientific facts before they were discovered by scientists, he could have easily told the biblical writers that the earth was a ball. Or he could have compared the earth to an orange or an apple. However, God never did this and did not tell us about the shape of our home, the earth. This is powerful evidence against the idea that God put modern science in the Bible.

The spiritual truth in Matthew 4:1-11 is clear. Jesus was tempted by the devil, but he did not bow down to Satan. Instead, Jesus replied to him by using a verse from the Bible, "Worship the Lord your God, and serve him only" (Deut. 6:13). And this spiritual truth is so important. Only God is worthy of worship.

To conclude, we can answer the question asked in the title of this chapter: Does the Bible have modern science? My answer is no. The Bible does not have modern science. Instead, the Bible has ancient science. This was the best science-of-the-day in ancient times. Therefore, God inspired the writers of the Bible by coming down to meet them at their level. God used their ancient understanding of nature as a vessel to transport life-changing spiritual truths. And these powerful truths are for all of us in every generation.

Ancient Science & the Genesis 1 Account of Creation

In this chapter, we discovered that there is an ancient science in the Bible. With this new knowledge, we can now have a better understanding of the first chapter of the Bible—the Genesis 1 account of creation in six days.

Before we look at the ancient science, it must be pointed out that the author of Genesis 1 was a brilliant writer! As you can see in Figure 3-6, he used a very structured writing style. Carefully arranged and constructed writing like this is called poetry. Genesis 1 features an ancient poetic framework called *parallel panels*.

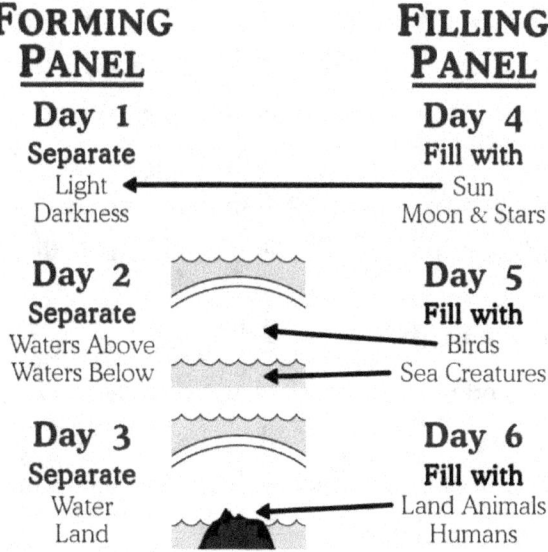

Figure 3-6. The Parallel Panels in Genesis 1

In the panel on the left, God forms and separates the boundaries of the universe. Then in the panel on the right, God fills the world by creating the heavenly sun, moon, and stars and the earthly animals and humans. The days in each panel are parallel to each other. In other words, the days line up across from one another. Let me explain.

On the first day of creation, God creates light and separates light from darkness. God then places the sun, moon, and stars in the firmament, on the fourth day of creation. The creation of light aligns with God filling the heavens with astronomical bodies that produce light.

On the second day of creation, God creates the firmament and then uses it to separate the waters above (heavenly sea) from the waters below (earthly sea). This forms an air space for birds to fly in. It also forms a body of water for sea creatures to swim in. On the fifth day of creation, God creates birds and sea creatures, and this aligns with what was made on day two.

On the third day of creation, God commands dry land to appear and he separates the land from the earthly sea. On the sixth day of creation, God makes land animals and humans. Once again, the days match up. Dry land is formed for land animals and humans to walk on.

I need to make a comment about the word day in Genesis 1. In the Bible, this word normally means a 24-hour day.

But it can refer to a period that is much longer. When the word day appears with a number, it means a 24-hour day. We see this in Genesis 1 with mention of the first day, the second day, etc. In addition, each creation day in Genesis 1 is described as a normal day, "There was evening, and there was morning—the first [second, etc.] day." Therefore, the days of Genesis 1 are regular days that are 24 hours long.

Two Important Lessons from the Ancient Science in Genesis 1
The ancient science in the Genesis 1 account of creation encourages us to think about this biblical chapter in new ways. There are two important lessons we can learn.

Lesson 1: Genesis 1 does not align with the natural world. For example, on the second day of creation, it states that God made a firmament to separate the heavenly sea from the earthly sea. On the fourth day of creation, it says that God placed the sun, moon, and stars in the firmament.

Of course, everyone knows that there is no hard firmament above our heads holding up a sea of water. And we also know that the sun, moon, and stars are not placed in a solid firmament. So, did God lie in the Bible? No, not at all.

God was meeting the writer of Genesis 1 and his readers at their level of understanding. God descended and allowed their ancient science to be used in the Bible. As I mentioned earlier, every Christian knows that God comes down to our level to meet us and to talk with us. God did the same thing when he inspired the writer of the Genesis 1 account of creation.

Lesson 2: Genesis 1 states that the universe and living creatures were made quickly and fully developed in their mature forms. As we have seen, the days of Genesis 1 are 24-hour days. According to Genesis 1, God made everything in only six days. However, ancient people did not have modern scientific instruments to tell them the world was billions of years old. And ancient people did not know about fossils. They were not aware that fossils are evidence that living creatures have evolved over millions of years.

The idea that everything was made quickly and fully developed is an ancient understanding of origins. It not only appears in Genesis 1, but also in the creation accounts of the nations surrounding ancient Israel. And this quick and complete creation of the universe and living creatures is also found in the creation accounts of ancient peoples around the world.

So again, we can ask the question. Did God lie in the Bible? And the answer is once more, no. Not at all. God came down to the level of the writer of Genesis 1 and used the best origins science-of-the-day to speak to him and his readers. At that time, everyone believed that the sun, moon, and stars, the plants and animals, and men and women were created quickly and fully mature.

But always remember, the ancient science in the Bible is a vessel that transports spiritual truths. In the Genesis 1 account of creation there are three main truths: (1) God created the universe and living creatures, (2) God created only humans in the Image of God, and (3) the creation is very good. It is these spiritual truths that change the lives of men and women. It is by believing these powerful truths we can develop a personal and loving relationship with our Creator.

CHAPTER 4
What Are the Different Choices for Origins?

In the first chapter of this book we discovered that there are more than only two choices for origins. We are not forced into choosing between either evolution or creation. It is possible that God could have used evolution to create the entire universe and all living creatures.

In this chapter, I will continue to show you that there are more than just two options for origins. In fact, there are five different choices for how the universe and living creatures were made: (1) young earth creation, (2) progressive creation, (3) evolutionary creation, (4) deistic evolution, and (5) atheistic evolution. The first three choices are accepted by Christians, and the last two are accepted by non-Christians.

It is important for you to know that there are many more views of origins than these five choices. I am presenting the five best-known positions so you can begin to figure out your own personal view on origins. As you go through this chapter, you might find that you accept different features from the different positions. It is totally acceptable to select and to combine parts of the five choices. If you do, then you will be developing one more view of origins—a sixth position on origins. And that's great! It means that you have moved beyond the evolution vs. creation debate.

Before we begin, I would suggest that you have a look at the chart that summarizes the five different positions of origins in Figure 4-5 on pages 98 and 99. This will give you an idea of where we are going in this chapter.

Young Earth Creation

Young earth creation claims that God created the entire universe and every living creature in only six 24-hour days about 6000 years ago. This view of origins is also known as six-day creation. Most people believe that young earth creation is *the* creationist position. They often assume that all Christians accept this view of origins. As I mentioned in Chapter 1, I once believed that *true* and *real* Christians had to be six-day creationists.

According to young earth creationists, the world has a plan and a purpose. They reject the idea that the universe is run only by luck and blind chance. Six-day creationists believe God planned the world specifically for us to live in. They say that God's ultimate purpose was for us to be in a personal and loving relationship with him. Young earth creationists also accept intelligent design. They experience the incredible beauty, complexity, and functionality in nature and claim that the creation shouts out to everyone that God exists.

Young earth creation does not believe the universe is billions of years old. And it does not accept the evolution of living creatures. Instead, this anti-evolutionary view of origins claims that God used miracles to create each plant, each animal, and each human in just six 24-hour days.

Six-day creationists read Genesis 1 in a strict literal way. They firmly believe that the days in this creation account are literal and actual 24-hour days. They claim that the Bible has modern science and that it tells us how God really created the world. Young earth creationists also believe that Genesis 1 offers us life-changing spiritual truths: (1) God is the Creator of the entire world, (2) men and women have been created in the Image of God, and (3) the creation is very good.

I have met many young earth creationists and they are wonderful Christians. They believe in a personal God and they enjoy a personal relationship with him. And they love the Bible. Most Christians today are six-day creationists. You might be interested to know that at my church most people are young earth creationists. But we know that our Christian faith is about Jesus, and only Jesus.

My Evaluation of Young Earth Creation

I was a young earth creationist for many years. As I learned more about how to read the Bible and studied science and fossils, I slowly moved away from this view of origins. Here are a few of the reasons why I am no longer a six-day creationist.

Previously, I pointed out that 98% of scientists in the United States accept the evolution of plants, animals, and humans over millions of years. We have all experienced the great blessings of science through medicine and engineering. So, should we believe that nearly every scientist in America is wrong about the origin of the world? Is that reasonable? What do you think?

We also saw that 40% of American scientists believe God answers prayer in a way that is more than merely subjective or psychological. In other words, these scientists experience that God is a personal God. Therefore, scientists are not all atheists. In addition, with 98% of US scientists accepting evolution, and 40% of them believing that God answers prayer, this would suggest that about 40% of American scientists believe God created the world through evolution. This tells us that believing in a Creator who used evolution is very reasonable.

As a young earth creationist, I would read the Genesis 1 account of creation in a strict and literal way. But I slowly realized this was not possible. The reason was because I discovered the Bible has ancient science. For example, the creation of the heavens in Genesis 1 does not match up with the physical world. As we saw, God created the firmament on the second day of creation. He then created the sun, moon, and stars and placed them in the firmament on the fourth day (See Figs. 3-3 and 3-6 on pages 51 and 65).

Of course, no one today thinks the dome of the sky is a hard and solid structure with the sun, moon, and stars in it. Therefore, the creation of the heavens in Genesis 1 is proof that we cannot read the Bible literally. In other words, the Bible does not have modern science as six-day creation claims.

Another problem I found with young earth creation deals with the pattern of fossils in the rock layers of the earth. One of the best features of science is that scientists make predictions, and then they test these predictions using evidence from nature to see if they are true. Figure 4-1 presents the fossil pattern expected by earth young earth creationists. Let me explain their scientific prediction for fossils.

Six-day creation claims that God miraculously created every type of living creature in only one week about 6000 years ago. Land plants and trees were made on the third day. Birds and sea creatures like fish and whales were created on the fifth day. Land animals such as amphibians, reptiles, and mammals were formed on the sixth day. And humans were also made on day six.

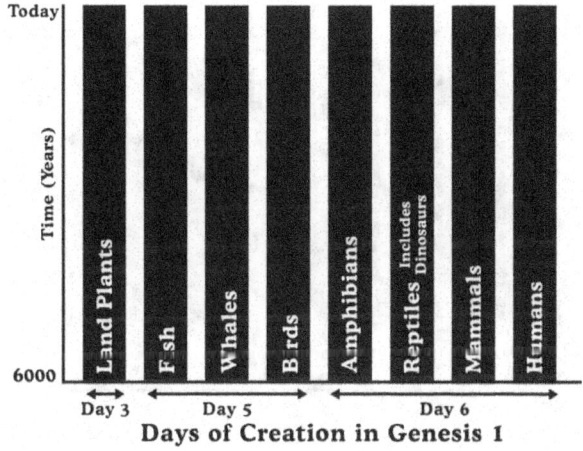

Figure 4-1. Fossil Pattern Prediction of Young Earth Creation

Each bar in Figure 4-1 represents a class of living creatures. And each of the bars is separate and not connected to the others. This is because young earth creation claims that no class of creature has ever evolved into another class of creature. For example, fish did not evolve into amphibians. The bars representing each class of living creatures extends from the bottom of the diagram when they were created 6000 years ago to the top of the diagram signifying today.

According to the fossil pattern prediction of six-day creation, at the very bottom of the fossil record there should be a layer of fossils with every kind of plant and animal that God made during the creation week in Genesis 1. If young earth

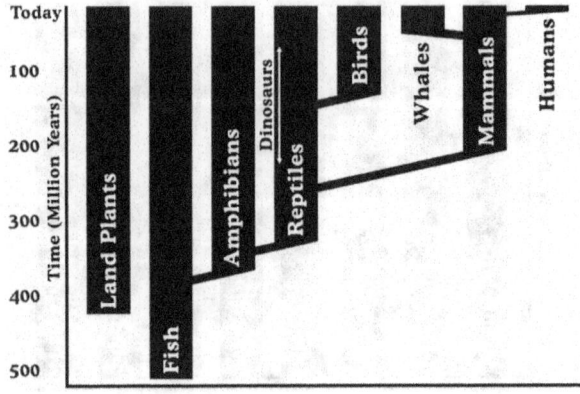

Figure 4-2. The Actual Fossil Pattern in Rock Layers of the Earth

creation is true, then the fossils of humans and dinosaurs (which are reptiles) should appear together in this lowest rock layer.

Figure 4-2 presents the actual pattern of fossils in the rock layers of the earth. It is important to emphasize that the order different plant and animal fossils appear has been known for over 200 years. No scientist today doubts this pattern. But it is obvious that the fossil pattern prediction of young earth creation does not match the scientific facts. And human fossils and dinosaur fossils are not found together. As Figure 4-2 shows, dinosaurs went extinct about 65 million years before we appeared on the earth.

There is another feature of the actual fossil pattern in Figure 4-2. Different classes of animals are connected by thin bars. These thin bars represent *transitional animals*. These creatures had features from the older group of animals *from* which they evolved. They also had features of the newer group *into* which they evolved. So, these animals could be called *in-between creatures*. Young earth creation claims there is no fossil evidence for transitional animals. But this is not true. Check out Appendix 2 where I present some fossils of animals evolving from one type of animal into a new type of animal.

To conclude, I must emphasize again that young earth creationists are wonderful Christians. We may disagree on how God created the world, but we firmly believe in the Creator. When I was a six-day creationist, there was one important idea I did not know. It was the idea that God came down to the level of the biblical writers and allowed them to use the science-of-the-day. Most young earth creationists are not aware the Bible has ancient science.

Progressive Creation

Progressive creation accepts the evolution of the universe, but it rejects the evolution of living creatures. This view of origins claims that the universe is about 14 billion years old and the earth around 4.6 billion years old. Progressive creationists also believe that God created different living creatures at different points in time. This view of origins is also called old earth creation and day-age creation. Progressive creation says that the days of creation in Genesis 1 are not 24-hour days, but periods of time that are millions of years long.

Like young earth creationists, progressive creationists believe that God created the world with a plan and a purpose. They do not believe that the universe is run only by luck and blind chance. Progressive creationists claim that God's main plan was to make a world for men and women. They also believe God created humans with the purpose to enjoy a personal and loving relationship with him. Progressive creationists also accept intelligent design. They claim that nature points everyone to God.

According to progressive creation, God created the world by using two different methods of creation. First, the Creator began the universe with the Big Bang. Over billions of years, the stars, planets, and moons slowly evolved into galaxies. In other words, God used natural processes to self-assemble these heavenly bodies. Second, progressive creationists believe that God used miracles to create every plant and animal, including humans. These anti-evolutionists claim that God entered the world at different times to create different living creatures.

Progressive creationists believe that the Genesis 1 account of creation lines up with modern science. They claim that the days of creation are millions of years long. These days are called *day-ages*. Progressive creationists say that the six day-ages in Genesis 1 align with the age of the universe at 14 billion years. They also claim that modern science matches up with the order that heavenly bodies and living creatures appear in Genesis 1.

According to the day-age interpretation of Genesis 1, God miraculously created land plants during the third day-age. He then waited millions of years and used more miracles to make the sun, moon, and stars in the fourth day-age. Similarly, he created birds and sea creatures on the fifth day-age. Finally, the Creator miraculously formed land animals and humans during the sixth day-age. In contrast to the strict literalism of young earth creation, progressive creationists accept general literalism when reading Genesis 1.

Many of my friends are progressive creationists and they are all marvelous Christians. We often have debates over how God created the world, but we completely agree that God really did create the world. We also realize that the main purpose of the Bible is to offer us spiritual truths. Like young earth creationists, progressive creationists believe the main spiritual truths in Genesis 1 are: (1) God is the Creator, (2) only humans have the Image of God, and (3) the creation is very good.

My Evaluation of Progressive Creation

There are many Christians who are attracted to progressive creation. This view of origins agrees with a lot of modern science. It accepts the Big Bang and that the universe is very old. Progressive creation also believes in the Bible. It claims that God gave the writer of Genesis 1 some basic facts of modern science. However, I find that there are some problems with this view of origins.

There is an inconsistency within progressive creation. On the one hand, it accepts the modern science of the evolution of the universe. Progressive creation claims that stars, planets, and moons self-assembled through natural processes. On the other hand, progressive creation rejects the evolution of living creatures. It states that God created plants and animals using miracles. But as we noted earlier, 98% of American scientists accept evolution. Are we to believe that scientists dealing with the evolution of the universe are all right, but scientists dealing with the evolution of living creatures are all wrong? That seems inconsistent to me.

Another difficulty with progressive creation is its understanding of the Bible. This view of origins believes the Bible has modern science. But as I have shown many times, the Bible has ancient science. The best example is the ancient astronomy in Genesis 1 (Figs. 3-3 and 3-6 on pages 51 and 65). I am sure you will agree that there is no way to align the firmament in the Bible with modern astronomy and the Big Bang.

The interpretation of the days of creation in Genesis 1 is another problem with progressive creation. It claims the days

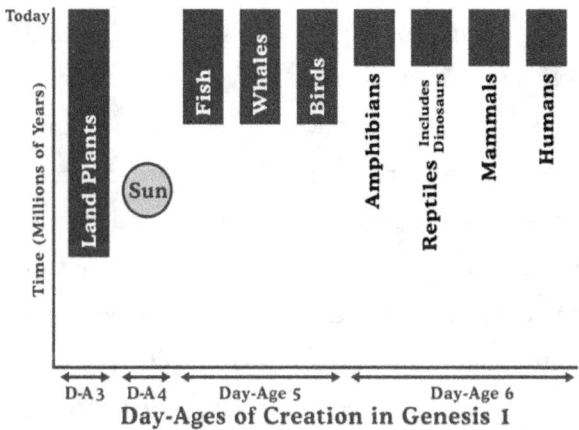

Figure 4-3. Fossil Pattern Prediction of Progressive Creation

are millions of years long. But as we saw in the last chapter, these days are 24 hours long. It's worth repeating the reasons. First, whenever the word day appears with a number in the Bible, it means a 24-hour day. The creation days in Genesis 1 are numbered. Second, each day ends, "There was evening, and there was morning." These can only be regular days. Genesis 1 does not support the idea that the days are day-ages that are millions of years long.

There are also scientific problems with progressive creation. Figure 4-3 presents the fossil pattern prediction of this anti-evolutionary view of origins. If progressive creation is true, then this is the order of fossils we should find in the rock layers of the earth. But as you can see, it does not match up with the actual fossil pattern discovered by modern science in Figure 4-2 on page 76.

Let me point out a few problems. According to progressive creation, birds and whales were created on the fifth day-age, while land animals were made on the sixth day-age. Therefore, birds and whales should be found millions of years before land animals. However, science has proven that land animals appeared 100s of millions of years before birds and whales. In fact, we know that birds evolved from small dinosaurs, and whales evolved from land mammals, as shown in the actual fossil pattern in Figure 4-2.

As I mentioned earlier, progressive creation rejects the evolution of living creatures. Therefore, the bars representing the creation of different animals in creation day-age five and day-age six are not connected to each other. Like young earth creation, progressive creation claims there are no transitional (in-between) animals. But again, this is not true. In Appendix 2 on page 116, I present an example of dinosaurs evolving into a bird. And on page 117, I show the fossils of a land mammal evolving into a whale.

Finally, according to progressive creation, the Genesis 1 account of creation states land plants were made on the third day-age. However, the sun was created on the fourth day-age. In other words, God created the sun millions of years *after* plants. But obviously, plants would not have survived for millions of years without light from the sun.

To conclude, the many problems with progressive creation offer us a valuable lesson: The Bible does not have modern science. Therefore, anyone who attempts to use the Bible like a book of science will fail. Instead, the Bible is a book of spiritual truths about who God is and who we are. The purpose of the Bible is to tell us how to develop joyful and loving personal relationships with God and with other people.

Evolutionary Creation

There is no doubt about it. When most people first hear the term evolutionary creation, it doesn't make any sense at all! It seems like a total contradiction. There are two reasons for this. First, most people think that evolution is atheistic and driven only by luck and blind chance. Second, they assume that creation means six-day creation. But if we use the definitions of evolution and creation that we introduced in Chapter 1, evolutionary creation makes perfect sense.

The most important word in the term evolutionary creation is the noun creation. Evolutionary creationists are first and foremost creationists. They believe in the Creator and that the entire world is his creation. The adjective evolutionary simply describes how God made the universe and living creatures. The Creator made all the laws of nature and every natural process, and through these the entire world self-assembled.

As I have just said, most people assume that evolution is driven only by luck and blind chance. Therefore, I need to clarify and emphasize some of the features about evolutionary creation:

- Evolutionary creationists are very dedicated Christians. They love God and they love the Bible. And they follow Jesus and his teaching.
- Evolutionary creationists believe that evolution has a plan and a purpose. It was God's plan to make a world for humans so they could have a personal relationship with him.
- Evolutionary creationists believe God caused the Big Bang. God set up the laws of nature and natural processes for the evolution of the universe and living creatures, including men and women.
- Evolutionary creationists believe in intelligent design. In fact, they claim that the incredible self-assembling process of evolution was intelligently designed and that it points us to God.
- Evolutionary creationists believe that God created humans in the Image of God. We are the only living creatures who have been blessed with this wonderful gift.
- Evolutionary creationists believe that all men and all women are sinners. God gave humans free-will.

But we often do not listen to God and we do not always obey his commands.
- Evolutionary creationists believe in miracles. They experience God's love and blessings in their life, including his answers to their prayers.
- Evolutionary creationists believe in the Bible. They claim that the main purpose of the Bible is to give us life-changing spiritual truths.

To summarize, evolutionary creationists firmly reject the atheistic belief that the world arose through evolution driven only by luck and blind chance. *Our existence is not a fluke or a mistake!* God planned and intelligently designed the process of evolution. He also maintained evolution during every second over billions of years. God intended for us to be here on earth. And God's main purpose for creating the world was for us to experience his love.

To be sure, most Christians find it very difficult to accept evolutionary creation. For many years, I struggled trying to make sense of this Christian view of origins. But as I pointed out in Chapter 1, there is a similarity between two different natural processes that helps us to understand evolutionary creation. Evolution and our own development in the womb resemble each other. Let me explain.

Every Christian believes that God created us in our mother's womb using developmental processes. No one thinks that God uses miracles to attach an entire arm or leg to a developing baby. *This is evidence that God creates using natural processes.* God intelligently designed the DNA in a fertilized human egg. It includes all the information that is needed for a person to gradually self-assemble in the womb during the nine months of pregnancy.

Similarly, the Creator planned the Big Bang and programmed it with the ability for the universe and living creatures to self-assemble over time. In other words, God intelligently designed evolution. He made the laws of nature and natural processes with all the information needed for the world to gradually evolve over 14 billion years. In particular, the Creator designed evolution to create humans and for us to become the most important living creatures of the evolutionary process.

There is another valuable similarity between evolution and our development in the womb. It deals with the appearance of the Image of God and sin with humans. Christians believe that during our own personal development, our Creator gave each of us the spiritual gift of the Image of God. We also believe that everyone becomes a sinner and is responsible for their own sins. In a somewhat similar way, during human evolution our pre-human ancestors became fully human when the Creator gave them the Image of God and free-will. And like each of us today, every one of these first men and women began to sin.

Evolutionary creation claims that God inspired the Bible. Like young earth creationists and progressive creationists, evolutionary creationists firmly believe the three central spiritual truths in Genesis 1: (1) God is the Creator of the entire universe and all living creatures, (2) only humans have been created in the Image of God, and (3) the world is very good.

However, evolutionary creationists disagree with young earth creationists and progressive creationists about what Genesis 1 says about the natural world. These two anti-evolutionary positions believe the Bible has modern science. Therefore, they assume that it is possible to make fossil pattern predictions, as we have seen in Figures 4-1 and 4-3. In contrast, evolutionary creation claims that the Bible has ancient science. As Figure 4-4 shows, this view of origins states that it is not possible to make fossil pattern predictions by using the Bible like a book of modern science.

NO FOSSIL PATTERN PREDICTION BASED ON THE BIBLE

The Bible has Ancient Science
It Cannot Be Used to Make
Scientific Predictions

Figure 4-4. Evolutionary Creation Rejects Fossil Pattern Predictions

Evolutionary creationists also point out there is ancient poetry in Genesis 1. This creation account is structured on a pair of parallel panels (Fig. 3-6 on page 65). But everyone knows that real events in the past do not follow poetic frameworks.

In fact, the creation of light on the first day of creation before the sun on the fourth day is proof of poetic freedom by the biblical writer. The reason the sun is created on day four is because it is in the filling panel of Genesis 1. Think about it for a moment. All ancient people knew that light comes from the sun because they experienced this every day! So, this is not a "contradiction" in the Bible. Instead, it is more evidence that God did not intend to reveal scientific facts in the Bible about how he made the world.

There is one final comment that needs to be made about evolutionary creation and the Bible. Many Christians assume that if this view of origins is true, then we should find evolution mentioned in the Bible. But this assumption is not correct. Evolutionary creationists claim that the Bible has ancient science. Therefore, the Bible has an ancient understanding of origins, and it will not have the modern science of evolution.

My Evaluation of Evolutionary Creation

The best feature of evolutionary creation is that it fully accepts *both* Christianity and modern science. This view of origins is very satisfying to people who experience God in a personal relationship and who know that God created the world through evolution. Evolutionary creation frees us from the assumption that there are only two choices for origins—either evolution or creation.

Evolutionary creationists enjoy a complementary relationship between the Bible and science. Truths from *both* the Bible and the natural world complete and strengthen each other. On the one hand, the modern science of evolution explains to us *how* God made this spectacular world and its countless examples of intelligent design. On the other hand, the Bible tells us *who* created the world—the God of Christianity.

One of the problems with evolutionary creation is that very few Christians hold this view of origins. Most Christians are either young earth creationists or progressive creationists. These anti-evolutionists believe that the Bible has modern science in Genesis 1. But evolutionary creationists suggest that we need to realize the Bible has ancient science. To accept this idea takes a lot of time. It took me many years before I was comfortable with the belief that God descended to the level of the writers of the Bible and allowed them to use the science-of-the-day.

I have noticed that nearly all older Christians reject evolution. They are the leaders of our churches, Sunday schools, and Christian schools. So, it is not surprising that young earth creation and progressive creation dominate Christian education. But I have also noticed a positive sign. Young people today are much more comfortable with evolution. I am convinced that when these young people become the next generation of leaders, evolutionary creation will be the most popular view of origins in Christianity.

The Bible & Real Events in History

Before ending this section on evolutionary creation, I need to make an important comment about historical events in the Bible. Evolutionary creationists claim that Genesis 1 has an ancient science of origins. As we saw at the end of the last chapter, ancient people believed the universe and living creatures were created quickly and fully formed. With this being the case, the Bible does not reveal the actual events in the past for how God created the world.

However, evolutionary creationists know that there are a lot of real people and real events mentioned in the Bible. In other words, the Bible includes a lot of accurate and real history.

For example, Abraham was a real person as stated in Genesis 11. There really was a King David in the city of Jerusalem during the 900s BC, as recorded in 2 Samuel 5. The Jews were taken as prisoners to Babylon in the 500s BC, as mentioned in Jeremiah 52. And there really was a man named Jesus who lived 2000 years ago. The Bible has many accounts from real people who heard the teachings of Jesus and saw his miracles, especially his resurrection after dying on the Cross.

Evolutionary creationists claim that the Bible does not tell us how the Creator made men and women. Again, creation accounts like Genesis 1 have an ancient understanding of origins. But evolutionary creationists firmly believe that the Bible records real events about the history of Jesus and his life. Jesus became a man and came to earth to teach us how to live a good life and how to love God and other people. And by dying on the Cross for our sins, Jesus showed us how much he loves us.

Deistic Evolution

The belief in a god who is not in a personal relationship with people is called *deism*. The deistic god is not involved in the lives of men and women. He never tells us who he is by inspiring written accounts like the Bible. This impersonal god never answers prayers and never uses miracles. In fact, it seems like the god of deism just doesn't care about his creation or us.

Deistic evolutionists believe that their god started the process of evolution with the Big Bang about 14 billion years ago. They also think that after causing this massive explosion, this god moved away from the universe and never came back to it. One way to picture deistic evolution is to see this creator as a god who winds up the world like a clock, and then he lets it run down on its own without ever entering it. In many ways, deism is like atheism. It is a world without a god.

By keeping their god outside the universe, deistic evolutionists completely reject the personal God of Christians. They do not accept the main belief of Christianity that God entered the world and became a man in the person of Jesus. And they completely reject the miracles of Jesus, like his resurrection after his death on the Cross. Deistic evolutionists also reject that God gave us spiritual truths in the Bible. They claim that the Bible is a fairy tale and it has no value for us today.

My Evaluation of Deistic Evolution

There is a major problem with an impersonal god who starts the evolution of the universe and then leaves never to return. This god rarely meets the spiritual needs of men and women. The deistic god leaves most people cold and uninspired. This god does not lead anyone to be involved in bettering the lives of other people. The reason is because the deistic god never enters the world to speak with us, challenge us, or motivate us.

In contrast to Christianity, deism never inspired a community of fellow deists, a teaching institution such as a university, or an organization to help people in need, like a hospital or a food bank. Deistic beliefs rarely transform the lives of people. But the teachings of Jesus have powerfully impacted men and women to dramatically change their lives. The church, the modern university, and the first hospitals and food banks were all inspired by Jesus calling Christians to do good things in the world and to care for other people.

Another problem with deistic evolution is that it seems very odd that an impersonal god would create personal creatures like us. Personal relationships are such an important part of everyone's life. Just think about how many emails and text messages you send to your friends every day. We are incredibly relational, and we need to be in personal relationships. That's who we are.

To me, it is more logical to believe that the Creator is a personal God who is in a personal relationship with men and women. The belief in the impersonal god of deistic evolution does not seem to be reasonable. Most people today and throughout history have experienced a personal God. And this is exactly the God of Christianity. He speaks to us through the Bible and in the small voice in our heart, and he even answers prayers with miracles.

Atheistic Evolution

Atheists do not believe God exists. They claim that evolution is driven only by luck and blind chance. Atheistic evolutionists say that the world does not have any plan or purpose. They also believe that humans are nothing but an accident and a fluke of evolution. Atheists claim that there is no ultimate (real) right or wrong, and that life is ultimately meaningless. Unfortunately, many people today assume that atheistic evolution is *the* evolutionist position and that all scientists hold this view of origins.

Atheistic evolutionists say that God is only an illusion. They believe that God is merely an invention of our imagination. Atheists also claim that intelligent design in nature is also an illusion. They say that our mind tricks us into believing that the world points to a Creator. And atheistic evolutionists often tell us that they are the smartest and most logical thinkers in the world. They also claim that religious people are irrational and ignorant.

Atheists believe that the Bible is only a fairy tale. They claim that the miracles of Jesus and his resurrection from the grave after his death on the Cross never happened. They say that miracles are based on the imagination and wishful thinking of Christians. Atheistic evolutionists also reject the spiritual truths in the Bible. They claim that these beliefs are nonsense, and no person should try living their life by following them.

My Evaluation of Atheistic Evolution

As I mentioned in Chapter 1, I became an atheist in college. Like most people, I assumed there were only two choices for origins—either atheistic evolution or young earth creation. When I discovered evidence for evolution in my biology classes, I thought I was making a very reasonable and logical decision to reject my Christian faith. But that was a very bad decision because I did not know the different positions on origins.

I returned to Christianity a few years after college. I slowly realized atheistic evolution had some serious problems. Let me share three reasons why I reject this view of origins.

First, not all scientists are atheists. Atheistic evolution is not the official view of origins in the scientific community. As we saw previously, a survey of American scientists discovered that 40% of them believe in a God who answers prayer. This survey also showed that 40% of these scientists believe in life after death. [12] It is not reasonable or logical to claim that a significant percentage of the best scientists in the United States are irrational and ignorant.

Second, most people today reject atheism. A survey of people throughout the world shows that only around 1% are atheists.[13] In the United States, less than 5% are atheists.[14] This survey also reveals that about 90% of Americans believe in God or a universal spirit. Therefore, it is not reasonable or logical for atheists to claim that nearly everyone today is trapped in an illusion because we believe in God.

Finally, the greatest problem with atheism is that atheists get rid of God and then they take his place! Atheists act like God because they decide what is right and wrong. But in my opinion, taking the place of God is arrogant. Maybe atheists are trapped in the illusion that there is no God.

Relationships between the Different Choices for Origins

Figure 4-5 on pages 98 and 99 presents the five basic positions on the origin of the universe and living creatures. This chart proves that there are more than just two choices for origins—either evolution or creation. Let's look at some relationships in the chart.

There are four types of creationists—young earth creation, progressive creation, evolutionary creation, and deistic evolution. They all believe in a Creator and that the world is his creation.

There are three different kinds of evolutionists—evolutionary creation, deistic evolution, and atheistic evolution. They all accept the evolution of the universe and living creatures.

Two positions on origins believe in God and accept evolution—evolutionary creation and deistic evolution. However, these positions have radically different religious beliefs. Deistic evolutionists believe in an impersonal god who never enters the world to meet us. They also say that the Bible is only a fairy tale and that it has no value for us today. In sharp contrast, evolutionary creationists believe in the personal God of Christianity. They enjoy a personal and loving relationship with him that includes the miraculous answers to their prayers. They also claim the Bible is inspired by God and offers valuable spiritual truths for our lives.

Figure 4-5 shows that there are three different Christian positions on origins—young earth creation, progressive creation, and evolutionary creation. They all believe that God created the world and that he is a personal God who acts through miracles in the lives of men and women. Of course, these three Christian views of origins disagree on how God created the world and how to read the Genesis 1 account of creation. Young earth creationists and progressive creationists claim the Bible has modern science. Evolutionary creationists believe that the Bible has ancient science and that Genesis 1 cannot be aligned with modern science.

In my opinion, knowing how God created the world is not necessary for being a Christian. As I mentioned earlier, I have met wonderful Christians who are young earth creationists, progressive creationists, and evolutionary creationists. So, I think that our differences about how God created the world should be just different points of view. And these differences should never divide us. We must always remember that we are united by our belief in Jesus and the fact that God loves each and every one of us.

	YOUNG EARTH CREATION	**PROGRESSIVE CREATION**
Plan & Purpose	Yes	Yes
Intelligent Design in Nature	Yes Points to God	Yes Points to God
Age of Universe	Young 6000 years	Old 14 billion years
Evolution of Living Creatures	No	No
God's Action in the Origin of the Universe & Living Creatures	Yes God uses miracles for living creatures & stars, planets, moons in only 6 days	Yes 1. God uses miracles for living creatures 2. God uses natural processes for stars, planets, moons
God's Action in the Lives of Men & Women	Yes Personal God uses miracles	Yes Personal God uses miracles
The Bible Has Spiritual Truths	Yes	Yes
The Bible Has Modern Science	Yes	Yes
Interpretation of Genesis 1	Strict Literalism Creation days 24 hours	General Literalism Creation days millions of years

Figure 4-5. Positions on the Origin of the Universe & Living Creatures

EVOLUTIONARY CREATION	DEISTIC EVOLUTION	ATHEISTIC EVOLUTION
Yes	Yes	No Plan & Purpose an illusion
Yes Points to God	Yes Points to God	No Design an illusion No God
Old 14 billion years	Old 14 billion years	Old 14 billion years
Yes	Yes	Yes
Yes God uses natural processes for living creatures & stars, planets, moons	Yes God uses natural processes for living creatures & stars, planets, moons	No No God God an illusion Natural processes are run by only luck & blind chance
Yes Personal God uses miracles	No Impersonal God No miracles	No No God God an illusion
Yes	No	No
No	No	No
1. Spiritual Truths 2. Ancient science 3. Ancient poetry	The Bible is a fairy tale	The Bible is a fairy tale

CHAPTER 5
Are You Ready to Make a Choice?

I hope you have enjoyed reading this book and that it has helped you to think about your personal view of origins. The main idea that I wanted to get across is that there are more than just two choices for origins—*either* atheistic evolution *or* creation in six literal days. As the title of this book suggests, I think everyone needs to move beyond the evolution vs. creation debate.

As we saw in the last chapter, there are five basic choices for origins: (1) young earth creation, (2) progressive creation, (3) evolutionary creation, (4) deistic evolution, and (5) atheistic evolution. The first three positions are accepted by Christians, and the last two by non-Christians. It is perfectly okay to select and combine features from the different choices. If you do, then you have created a sixth position on origins. And this moves you beyond the evolution vs. creation debate.

One of the advantages of studying the origins debate is that it encourages us to think hard about what we believe. In other words, this is a great topic for developing critical thinking skills. It is usually easy to say *what* we believe. But it is much tougher to say *why* we believe. By showing you the many different choices for origins, you can think about which ones seem best to you. And you can also start thinking about the reasons for your choices.

In my opinion, there are two main topics that you need to consider when developing your personal view of origins. First, what about intelligent design? Does the beauty, complexity, and functionality in nature point to God? As a scientist who studies the development and evolution of teeth and jaws, I am blown away by the marvelous creation! To believe that the universe and living creatures evolved because of only luck and blind chance seems impossible to me. There has to be an intelligent mind that planned our self-assembling world. I believe this is the Mind of God.

Second, what about the passages in the Bible that refer to nature? Does the Bible have modern science? Or does it have ancient science? For most Christians, the idea that there is an ancient understanding of the natural world in the Bible is a new idea. And it's very challenging. But as I have pointed out many times, Christians know that God meets us wherever we are in our spiritual voyage in order that we may understand him. So, did God meet the ancient writers of the Bible at their level of understanding and use their ancient science?

To ask this second question is threatening to many Christians. But if we are going to be good critical thinkers, we need to ask whether God allowed the writers of the Bible to use the science-of-the-day. God has given us a brilliant mind and he wants us to use it. I am completely convinced that when we ask tough questions like this, it pleases our Creator. Remember what Jesus commanded us to do. We need to love God "with all our mind" (Matt. 22:37).

There is one piece of advice that I always give to my students at the end of my course on science and religion. I'll suggest it to you as well. There is no rush to figure out what your views are on origins. It's okay not to know and not to have an opinion on this topic. Take your time. Be comfortable with your decisions. But never stop thinking. Always keep moving forward.

And if I can make one more suggestion. Share your thoughts on origins with your family and friends. Learning from others is an important part of developing our personal beliefs. My students have had a huge impact in shaping my views on science and religion. They have opened my eyes to blind spots in my thinking and things I did not see.

To encourage you to share your view on origins with others, Appendix 1 has a pair of surveys with the five basic positions on origins. Copy and cut them apart. You could use these in group discussions. I use this survey in all my classes. My students have told me that calculating the percentages for each position in the class is helpful to them. Christian students are always surprised and encouraged to discover that there are more Christians who accept evolution than they thought.

Evolution & Creation in a Complementary Relationship

Now that we have moved beyond the evolution vs. creation debate, I can propose a way to think about evolution and creation in a complementary relationship, as shown in Figure 5-1. As we saw previously, the word complementary means to make something complete, whole, or perfect. A complementary relationship is a two-way relationship. In a complementary relationship between two people, each person adds something that the other person does not have. Together, they inform, improve, and strengthen one another.

CREATION
Spiritual Truth
Universe & Living Creatures
Created by God

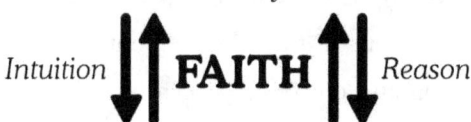

Intuition **FAITH** *Reason*

EVOLUTION
Scientific Fact
Universe & Living Creatures
Evolved over Time

Figure 5-1. Evolution & Creation in a Complementary Relationship

A complementary relationship between evolution and creation is also a two-way relationship. On the one hand, the spiritual truth that the universe and living creatures are the creation of the Creator informs scientists about *who* caused the Big Bang and the process of evolution. The downward arrows in Figure 5-1 show that this religious belief improves our understanding of the ultimate meaning of evolution.

On the other hand, the scientific fact that the world evolved over billions of years informs religious people about *how* God created the world. The upward arrows in Figure 5-1 indicate that evolution improves our knowledge about the Creator and the creation. God used evolution and his world is an evolutionary creation. Together, the religious belief in creation and the scientific theory of evolution complete and strengthen each other.

I need to emphasize that a step of faith (or intellectual leap) always happens in this complementary relationship between evolution and creation, as seen in the center of Figure 5-1. Of course, we use our intuition and our reason in thinking about origins. But in the end, it takes faith to believe that God created the universe and living creatures through evolution.

God's Acts of Creation & the Shooting Billiards Analogy

Let me offer an analogy to explain why I think God using evolution to create is incredibly amazing. Try to imagine that

God's acts of creation are like the stroke of a cue stick in a game of billiards. Divide and label the billiard balls into three groups using the words heavens, earth, and living creatures. And let the eight ball—the most important ball in billiards—represent humans.

With this analogy, we can picture the God of young earth creation and his many miracles to create the universe and living creatures. This Creator makes single shot, after single shot, after single shot, etc. He does not miss a shot. And he clears all the balls off the table into the billiard pockets. To be sure, that's quite impressive.

But an evolutionary creationist sees God using only one stroke of his cue, representing the Big Bang. This single shot is so powerful and so precise that all the balls are sunk! And not only that, they even drop in order: first, the balls labeled the heavens, next the earth, and then the living creatures. The eight ball representing humans is the last ball to fall. God then pulls this ball out of the billiards pocket and holds it to his heart to indicate he is in a personal relationship with men and women.

Isn't the Creator who uses just one single stroke of the cue to sink all the balls so much more spectacular than the God of young earth creationists who takes shot, after shot, after shot? For me, God's tremendous power and planning is best seen with him creating the world through evolution. Just think about it. With only one act of creation, God caused the Big Bang. The laws of nature emerged and then the entire world self-assembled through evolution. I can't think of a more mind-blowing example of intelligent design.

The Evolution & Creation of the Brain

To further explain the relationship between evolution and creation, let's consider the most complex structure known in the universe—the human brain. We have about 100 billion brain cells and there are roughly 100 trillion connections between them. Fully written out, that's 100,000,000,000 cells and 100,000,000,000,000 connections. When we consider that our Milky Way galaxy has about 1 billion stars, the complexity of the brain is overwhelming. Are we to believe that our brains evolved only because of luck and blind chance? I find that belief completely unreasonable.

Instead, I believe that God planned and intelligently designed evolution to create our brain. The Bible states that we have been created in the likeness and Image of God (Gen. 1:26-27). I believe this is true. So, if the Creator of this marvelous world made us somewhat similar to him, this means that we must have awesome creative gifts. And this is also true.

Think about it. With our incredibly creative brains that God has given us, we have put rovers on Mars. We have made vaccines in a record time to protect us from Covid-19.

We have created computers so we can have Zoom conversations anywhere on earth. Are we to believe that our brain is just a fluke of nature caused only by luck and blind chance? Again, I find that belief completely unreasonable.

The creation and evolution of our marvelous brain can be viewed in the light of the complementary relationship in Figure 5-1. By taking a step of faith downward, my belief that God created everything informs and completes what science says about the evolution of the brain. It tells me that God designed the evolutionary process that created our wonderful brain. The scientific study of the brain demonstrates its spectacular complexity and incredible functionality as seen through its countless achievements, such as modern science. In taking a step of faith upward, I use this scientific evidence to strengthen my belief in both intelligent design and the Creator. Indeed, our brain has been created in the likeness and image of the Mind of God!

A Few Final Thoughts

I am sure you have figured it out that I am an evolutionary creationist. I want you to know that believing in this view of origins did not come quickly or easily. As I mentioned in earlier chapters, I began my spiritual voyage as a young earth creationist. Then for over ten years, I struggled with the interpretation of the creation accounts in the Bible and I also explored the scientific evidence for evolution. Eventually, I accepted evolutionary creation.

During this period of trying to make sense of origins, I was not aware of many Christians who accepted evolution. And I did not know of one book on evolutionary creation. This is one of the reasons why I wrote this book. I want you to be aware of this view of origins. Evolutionary creation is a choice you can make as a Christian.

Another thing that I suspect you have figured out is the image on the cover of this book. Yes, it is based on DNA, the double-stranded molecule of life. This image also features the two main ways that God speaks with us.

Green is often viewed as the color of life. The green strand of DNA with the arrow pointing upward indicates that nature points us to God. Purple is the color of royalty for kings and queens. The purple strand of DNA and the downward arrow refer to our Creator the King of the Universe coming down from heaven to meet us at our level on earth. Most importantly, God descended and became a human in the person of Jesus to say to us and to show us how much he loves us.

In closing, I want to thank you for taking the time to read my book. Let me end with one last question and one last answer. Is it possible to have a positive and productive relationship between the scientific theory of evolution and the religious belief in creation? For me, the answer is an absolute YES! As a Christian who is also a scientist, my belief in the Creator and my acceptance of evolution offer me an informed and complete understanding of the world. But before I came to believe that the Creator created an evolving creation, I first needed to move beyond the evolution vs. creation debate.

ACKNOWLEDGEMENTS

I am grateful to many people who have assisted and encouraged me in the writing of this book. My teaching assistant Anna-Lisa Ptolemy worked tirelessly and meticulously on editing numerous versions of the manuscript. Many thanks to my collaborators in the Faculty of Education at the University of Alberta, Scott Key, Norma Nocente, and Colleen Rizzoli. Our newly developed professional education course for teachers inspired me to write this book. It was again a pleasure to work with Caleb Poston and Duane Cross at McGahan Publishing House. Thank you to Andrea Dmytrash, Danica Wolitski, and Kenneth Kully for their artwork. Others who have contributed to my work include Chris Barrigar, Michael Caldwell, Keith Furman, Sy Garte, Brian Glubish, Wendell Grout, David Haitel, Reagan Haitel, Pliny Hayes, Kristine Johnson, Bob Lamoureux, Viet Nguyen, Isabel Ptolemy, Randy Isaac, Anita and Paul Seely, Corwin Sullivan, Jennifer Swainson, and Madison Trammel. And I am thankful for the support of President Shawn Flynn and Dean Matthew Kostelecky at my college.

BIOGRAPHY

Dr. Denis O. Lamoureux is Professor of Science and Religion at St. Joseph's College in the University of Alberta. He is also a research scientist in paleontology. Lamoureux holds three earned doctoral degrees—dentistry, theology, and biology. He has written several books on the origins debate, including *The Bible & Ancient Science: Principles of Interpretation* (2020). Denis worships at Hope City Pentecostal Church in Edmonton, Alberta.

Appendix 1
Group Discussion Surveys

POSITION ON ORIGINS
No Names Please

Check off (✓) the position that best describes your view

- [ ] **Young Earth Creation**
  God created in 6 days 6000 yrs ago
- [ ] **Progressive Creation**
  God created living creatures at different times over billions of years
- [ ] **Evolutionary Creation**
  Personal God created using evolution
- [ ] **Deistic Evolution**
  Impersonal God created using evolution
- [ ] **Atheistic Evolution**
  No God. Evolution by luck & blind chance

Comments on Back
Any stories? Any questions?
- [ ] I **DO NOT** want my comments used publicly

POSITION ON ORIGINS
No Names Please

Check off (✓) the position that best describes your view

- [ ] **Young Earth Creation**
  God created in 6 days 6000 yrs ago
- [ ] **Progressive Creation**
  God created living creatures at different times over billions of years
- [ ] **Evolutionary Creation**
  Personal God created using evolution
- [ ] **Deistic Evolution**
  Impersonal God created using evolution
- [ ] **Atheistic Evolution**
  No God. Evolution by luck & blind chance

Comments on Back
Any stories? Any questions?
- [ ] I **DO NOT** want my comments used publicly

Copy & Cut

Appendix 2
Transitional Fossils

There are thousands of transitional fossils. In this short appendix, only 3-4 examples are presented for each major transition. But the pattern of evolutionary change is obvious.

Early Amphibian   365 mya

Fish with Limb-Like Fins   375 mya
   This fish could crawl on land

Lobe-Finned Fish   385 mya
   Lobe fins have bones similar to limbs (See next figure)

Figure 1. Fish-to-Amphibian Evolution
   See Fig. 2 for the bones in fins and limbs.
      mya: millions of years ago   Image Credits, p. 118.

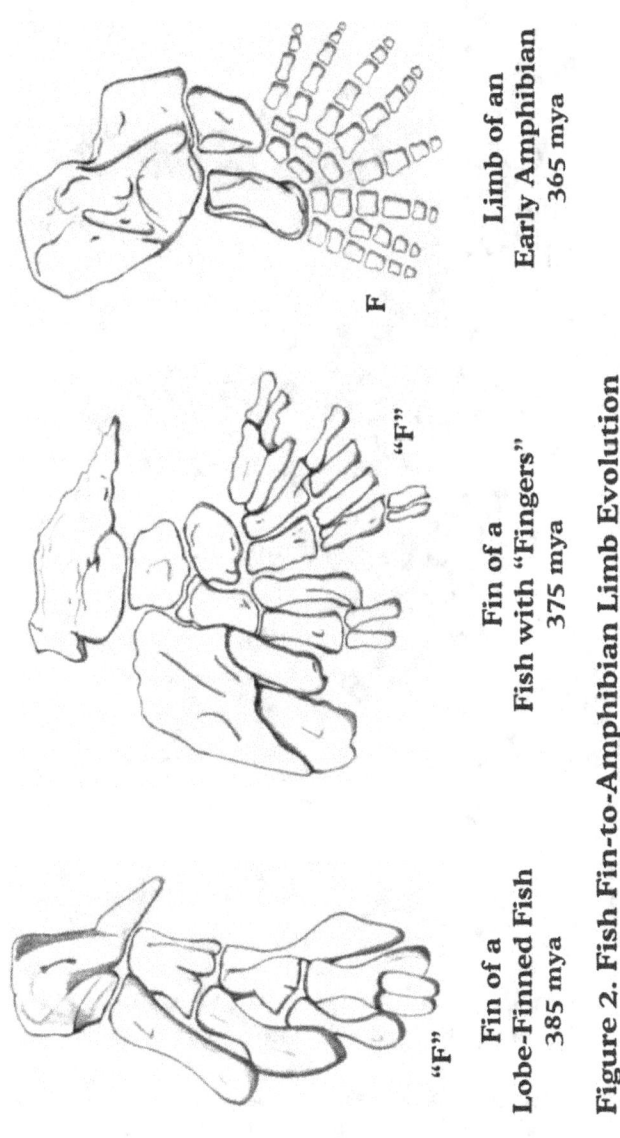

**Fin of a
Lobe-Finned Fish
385 mya**

**Fin of a
Fish with "Fingers"
375 mya**

**Limb of an
Early Amphibian
365 mya**

Figure 2. Fish Fin-to-Amphibian Limb Evolution

Early amphibians often had more that five fingers because they evolved from fish with many finger-like bones, such as the fish above with 9 "fingers." The amphibian above had 8 fingers.
F: fingers "F": finger-like bones mya: millions of years ago

Image Credits, p. 118-119.

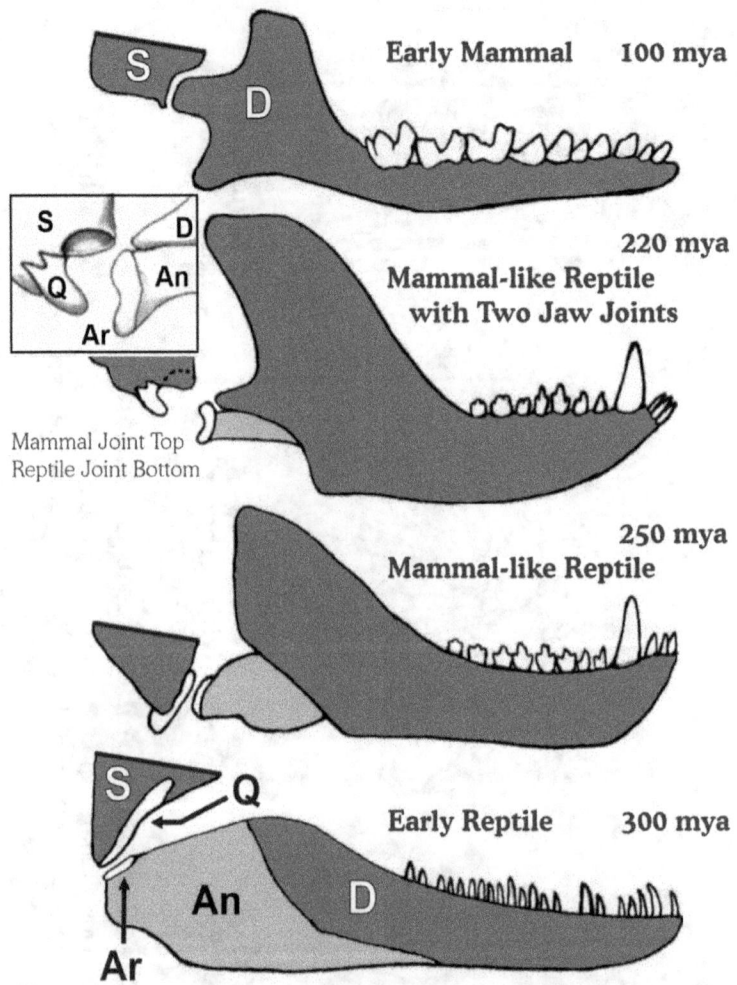

Figure 3. Reptile-to-Mammal Jaw Evolution
In the evolution of reptiles-to-mammals, the bones of the jaw joint changed. During the transition, some mammal-like reptiles had both types of jaw joints. Eventually, the reptile joint was lost in mammals.
Ar: Articular Q: Quadrate D: Dentary S: Squamosal An: Angular Credits, p. 119.

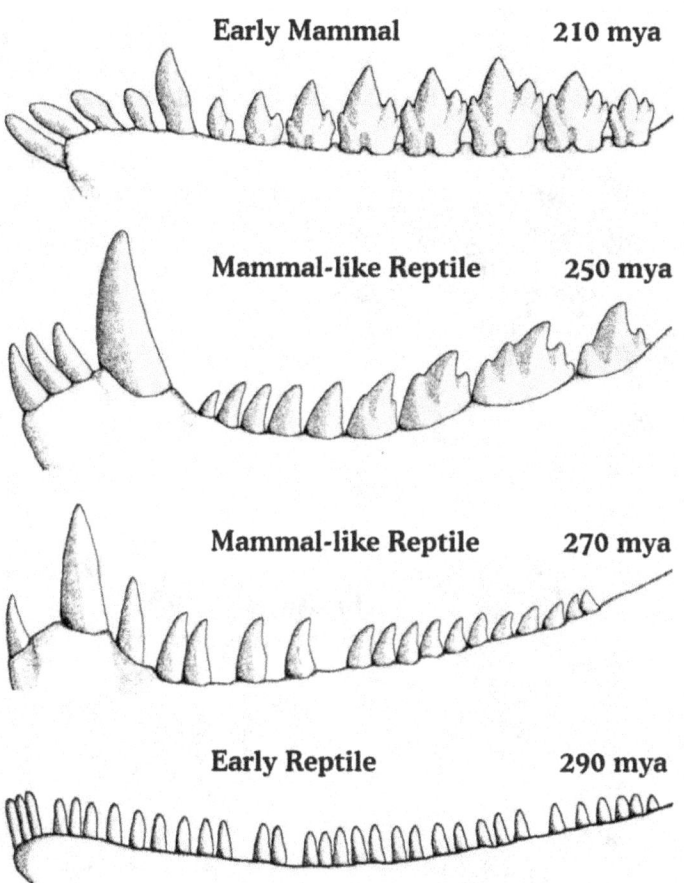

Figure 4. Reptile-to-Mammal Tooth Evolution
Reptiles have teeth shaped like cones. These teeth are not useful for grinding up food. Mammal-like reptiles have a large killer tooth (like the canine in dogs) and some back teeth are beginning to evolve points. Mammals have sharp points on their back teeth. This allows them to chew up their food and to get more nutrition.
mya: millions of years ago   Image Credits, p. 119.

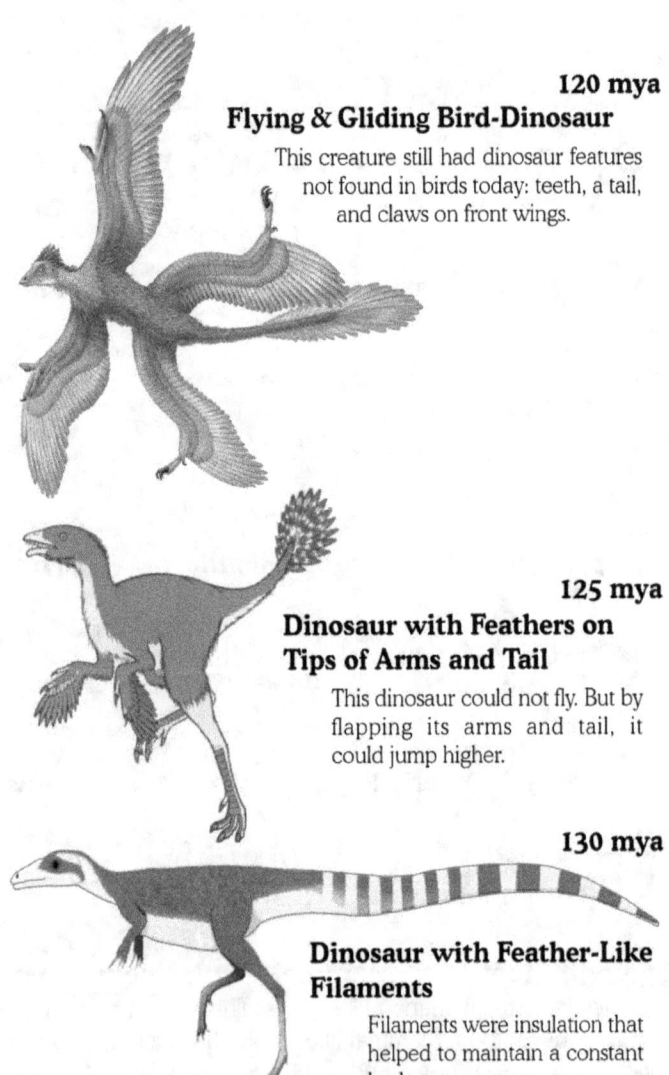

120 mya
Flying & Gliding Bird-Dinosaur

This creature still had dinosaur features not found in birds today: teeth, a tail, and claws on front wings.

125 mya
Dinosaur with Feathers on Tips of Arms and Tail

This dinosaur could not fly. But by flapping its arms and tail, it could jump higher.

130 mya
Dinosaur with Feather-Like Filaments

Filaments were insulation that helped to maintain a constant body temperature.

Figure 5. Dinosaur-to-Bird Evolution
mya: millions of years ago   Image Credits, 119-120.

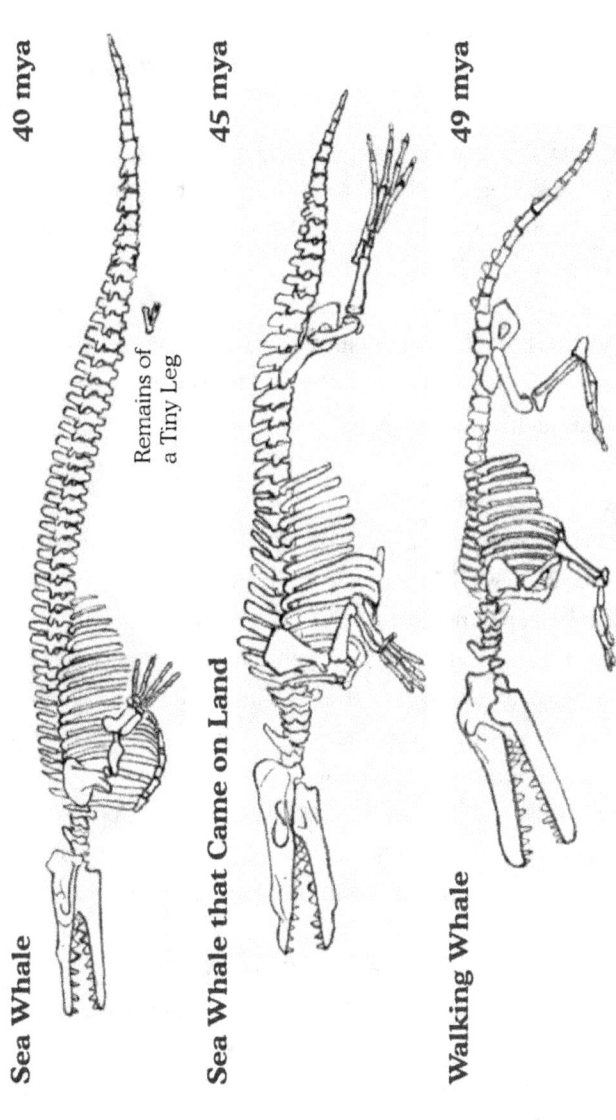

Figure 6. Land Animal-to-Whale Evolution Image credits, p. 120.
Early whales had huge feet and this improved their ability to swim. As whales evolved, the back leg became smaller. Even today, whales have the remains of a tiny leg. mya: millions of years ago

APPENDIX IMAGE CREDITS

Fish-to-Amphibian Evolution

• Early Amphibian. Tamura, Nobu (21 January 2007) *Acanthostega gunnari*. Own work, CC BY 2.5. Retrieved June 24, 2018 from: https://commons.wikimedia.org/wiki/File:Acanthostega_BW.jpg. Permission is granted to copy, distribute and/or modify this document under the terms of the GNU Free Documentation License, Version 1.2 or any later version published by the Free Software Foundation.

• Fish with Limb-like Fins. Tamura, Nobu (22 January 2007) *Tiktaalik rosae*. Own work, CC BY 2.5. Retrieved June 23, 2018 from: https://commons.wikimedia.org/wiki/File:Tiktaalik_BW.jpg. Permission is granted to copy, distribute and/or modify this document under the terms of the GNU Free Documentation License, Version 1.2 or any later version published by the Free Software Foundation.

• Lob-Finned Fish. Tamura, Nobu (2 May 2007) *Eusthenopteron foordi*. Own work, CC BY 2.5. Retrieved June 23, 2018 from: https://commons.wikimedia.org/w/index.php?curid=19461628 . Permission is granted to copy, distribute and/or modify this document under the terms of the GNU Free Documentation License, Version 1.2 or any later version published by the Free Software Foundation.

Fish Fin-to-Amphibian Limb Evolution

• Fin of a Lobe-Finned Fish. Dmytrash, A. (2009) Eusthenopteron. Based on, adapted, and redrawn from: Robert L. Carroll, *Patterns and Processes of Vertebrate Evolution* (New York, NY: Cambridge University Press, 1997), 233. Used with permission.

• Fin of a Fish with "Fingers." Dmytrash, A. (2009) *Sauripterus*. Based on, adapted, and redrawn from: Edward B. Daeschler and Neil Shubin, "Fish with Fingers?" *Nature* 391 (January 8, 1997), 133. Used with permission.

• Limb of an Early Amphibian. Dmytrash, A. (2009) *Acanthostega*. Based on, adapted, and redrawn from: M. I. Coates, J. E. Jeffrey, and M. Rut,

"Fins to Limbs: What the Fossils Say," *Evolution and Development* 4 (2002): 392. Used with permission.

Reptile-to-Mammal Jaw Evolution

• Early Mammal Jaw. Dmytrash, A. (2009) *Dautestes*. Based on, adapted, and redrawn from: Kenneth D. Rose, *The Beginning of the Age of Mammals* (Baltimore: John Hopkins University Press, 2006), 92. Used with permission.

• The Two Mammal-like Reptile Jaws and Early Reptile Jaw. Dmytrash, A. (2009) Probingognathus (upper) *Thrinaxodon* (lower). Based on, adapted, and redrawn from: Robert L. Carroll, *Vertebrate Paleontology and Evolution* (New York: W. H. Freeman, 1988), 366, 382, and 390. Used with permission.

Reptile-to-Mammal Tooth Evolution

• The Teeth of Early Mammal, the Two Mammal-like Reptiles, and Early Reptile. Barr, B. (2008). From top to bottom: *Morganucodon, Cynognathus, Dimetrodon, Protorothyris*. Based on, adapted, and redrawn from: Robert L. Carroll, *Vertebrate Paleontology and Evolution* (New York: W. H. Freeman, 1988), 196, 365, 386, 406, 408. Used with permission.

Reptile-to-Bird Evolution

• Flying and Gliding Bird-Dinosaur. Portia Sloan Rollings (2013) *Microraptor*. Used with permission.

• Dinosaur with Feathers on Tips of Arms and Tail. Conty (2021) *Caudipteryx*. Retrieved May 3, 2021 from: https://commons.wikimedia.org/wiki/File:Caudipteryx_0988.JPG. This file is licensed under the Creative Commons Attribution-Share Alike 3.0 Unported license.

• Dinosaur with Feather-like Filaments. Smithwick, F.M., Nicholls, R., Cuthill, I.C., and V., Jakob, (2017) *Sinosauropteryx*. Source http://www.cell.com/current-biology/fulltext/S0960-9822(17)31197-1. Retrieved May 24, 2018 from: https://commons.wikimedia.org/wiki/File:Sinosauropteryx_color.jpg. This file is licensed under the Creative Commons Attribution 4.0 International license.

120 | Beyond the Evolution vs. Creation Debate

Land Animal-to-Whale Evolution
• Walking Whale, Sea Whale that Came on Land, and Sea Whale. Dmytrash, A. (2009). From top to bottom: *Durodon, Rodhocetus, Ambulocetus*. Based on, adapted, and redrawn from: Kenneth D. Rose, *The Beginning of the Age of Mammals* (Baltimore: Johns Hopkins University Press, 2006), 282. Used with permission.

NOTES

[1] Ninety-eight percent of American scientists accept that "humans and other living things have evolved over time." No Author, "Elaborating on the Views of AAAS Scientists, Issue by Issue" Pew Research Center: Science and Society (23 July 2015). Accessed March 20, 2021: http://www.pewresearch.org/science/2015/07/23/elaborating-on-the-views-of-aaas-scientists-issue-by-issue/.

[2] Human Development in the Womb. Dmyrash, A. (2021). Own artwork. Used with permission.

[3] I will use the New International Version of the Bible (2011) for biblical passages in this book. But occasionally I use my own translation to make the verse more understandable. In Psalm 139:14, I have replaced the common translation of "fearfully" with "amazingly." The term "fearfully" carries too much negative baggage.

[4] Bacterial Flagellum. Kully, Kenneth (2014). Based on, adapted, and redrawn from: D. Voet and J. G. Voet, *Biochemistry*, 2nd ed. (New York: Wiley, 1995), 1259. Used with permission.

[5] Edward J. Larson and Larry Witham, "Scientists Are Still Keeping the Faith," *Nature* 386 (April 3, 1997): 435.

[6] The Big Bang. Lamoureux, D.O. (2021). Based on, adapted, and redrawn from: A. Fraknoi. D. Morrison, and S. Wolf, *Voyages through the Universe* (Fort Worth, TX: Saunders, 1997), II: 578.

[7] Paul W. C. Davies, *God and the New Physics* (London, UK: Penguin, 1983), 179.

[8] Hugh Ross, *Improbable Planet: How Earth Became Humanity's Home* (Grand Rapids, MI: Baker Books, 2017).

[9] Simon Conway Morris, *Life's Solution* (Cambridge, UK: Cambridge University Press, 2003).

[10] See endnote 3. I will use the singular form for the dual Hebrew word "heavens."

[11] Babylonian World Map. Lamoureux, D.O. (2021). Based on, adapted, and redrawn from: Map of the World [Online image of a Babylonian World Map, British Museum Image Number: 92687]. Retrieved January 23, 2019 from: http://www.britishmuseum.org/research/collection_online/collection_object_details/collection_image_gallery.aspx?partid=1&assetid=32436001&objectid=362000. Used with permission of The British Museum under a Creative Commons Attribution-NonCommercial-ShareAlike 4.0 International license. © Trustees of the British Museum,

[12] Edward J. Larson and Larry Witham, "Scientists Are Still Keeping the Faith," Nature 386 (April 3, 1997): 435

[13] See "Secular/Nonreligious/Agnostic/Atheist" in "Major Religions of the World Ranked by Number of Adherents," accessed June 4, 2015, www.adherents.com/Religions_by_Adherents.html.

[14] "Not All Nonbelievers Call Themselves Atheists," Pew Research Center (April 2, 2009), accessed June 4, 2015, www.pewforum.org/2009/04/02/non-all-nonbelievers-call-themselves-atheists.

INFORMATION ON BACK COVER

A 2011 Barna Group survey reveals that 59% of young people "disconnect either permanently or for an extended period of time from church life after age 15." The issue of science is one of the reasons why they leave the church. This study records that 25% of them perceive that "Christianity is anti-science," and 23% have "been turned off by the evolution-versus-creation debate." No Author, "Six Reasons Young Christians Leave Church," Sept 28, 2011. Retrieved June 24, 2018 from: https://www.barna.org/teens-next-gen-articles/528-six-reasons-young-christians-leave-church.

www.ingramcontent.com/pod-product-compliance
Lightning Source LLC
Chambersburg PA
CBHW050326120526
44592CB00014B/2075